Bird Tales

A Lifetime Pursuit

George C. Wood

Bird Tales
A Lifetime Pursuit

Except where noted, all bird photographs by the author.

ISBN-13: 978-0-9800182-4-0
ISBN-10: 0-9800182-4-2

georgecwood@comcast.net

My father told me many years ago that if I wanted to leave a legacy, I should procreate and/or write a book. I elected to do both but not at the same time.

My goal here is to condense over fifty years of birdwatching adventures into an entertaining and informative narrative for both birders and non-birders.

Join me as I relive the special moments and milestones in my lifetime pursuit of seeing 800 bird species in America.

Acknowledgments

I am forever grateful to my parents, Jeff Dingle, and the Severinghaus family for introducing me to a lifelong hobby.

No matter where I am or what I am doing, I am never bored, because birds are everywhere.

Of course, I want to thank my family, friends, and especially my loyal blog followers for providing support and motivation to keep birding and pursuing my goal. This adventure has been a marathon, not a sprint.

My sister-in-law Myra Wood deserves a special shout-out for professionally guiding me through the self-publication process. And she deserves credit for locating my copy-editor, Lisa White, who is amazing.

And finally, to my wife, Nina Weisbord, I appreciate your understanding, patience, and unconditional love.

Thank you, all.

Front cover: The ***Great Kiskadee*** *is a large, aggressive flycatcher considered a specialty of Southern Texas and the Rio Grande Valley. It is an omnivore that captures insects in flight, devours small rodents, and even eats fish in shallow water.*

Contents

1. Brag Card

At age thirteen, I saw the rare California Condor in the wild, which was an experience of a lifetime. The federal government had recently placed the condor on the endangered species list, as fewer than fifty of these majestic birds with a ten-foot wingspan existed in the world. They could be found only in the Los Padres National Forest north of Los Angeles.

I spent my eighth-grade spring vacation visiting my cousins in Palos Verdes, California. My uncle Wistar was a biology teacher at the Chadwick School, and he kindly took the time to drive us to the summit of Mount Pinos, where we waited in vain for several hours to observe these soaring birds. Three days later we repeated this long day trip, and this time we were fortunate enough to observe three adult condors flying over our heads. I will never forget that special moment. Our perseverance paid off.

America's awareness of nature's frailties had been elevated a year earlier with the celebration of our initial Earth Day on April 22, 1970. Cornell University biologists captured the few remaining California Condors in the wild in the 1980s. By 1987 this species was declared extinct in the wild.

California Condor *(January 2013, Bitter Creek National Wildlife Refuge, California)*

Thankfully, a successful captive-breeding program that is ongoing today has reintroduced a couple of hundred condors in northern Arizona, southern Utah, and coastal mountain areas of central and southern California. The California Condor is still a critically endangered species and one of the rarest birds on earth, as a combined total of only 500 or so individuals live in the wild or in captivity.

My brag card among other birders is the fact that I saw the California Condor in its wild habitat *before* captive breeding. But my interest in the natural world and, specifically, birds, had begun several years before that memorable day in 1971.

2. The Influencers

George's parents: Anne and Howard Wood

My parents, Anne and Howard Wood, were interested in flora and fauna. Dad was a physician and an expert in *Dendrobium* orchids. We used to tease him that he was a "part-time doctor." Every house he lived in had included a greenhouse.

As a youngster, in order to spend time with him, I'd accompany Dad on jaunts to New Jersey's Pine Barrens. He liked birds, as well, so trips to Brigantine and Cape May, New Jersey, Bombay Hook, Delaware, and Hawk Mountain, Pennsylvania facilitated the development of a slight interest in birds at an early age.

My father was red-green color blind. When it came to identifying bird species, especially warblers, he taught me to focus on listening to sounds rather than try to distinguish colors. To this day, that is the best strategy for learning to discern songbirds.

Jeff Dingle with a portrait of his grandfather,
Dr. Leslie Severinghaus, fourth head of The Haverford School

During spring vacation of fifth grade, my friend Jeff Dingle invited me to visit his grandparents, the Severinghauses, in Coconut Grove, Florida. I immediately accepted, as I thought we'd spend most of the time watching exhibition baseball games and simultaneously get a suntan. Dr. Severinghaus, a former head of school of my alma mater, The Haverford School, had other plans.

Each day we took car trips to watch birds in places such as the Everglades, Lake Okeechobee, and the Florida Keys. With encouragement from Dr. Severinghaus, both Jeff and I relished the challenge of quickly and accurately identifying whatever birds we saw. I also attempted to capture photographs with my trusty Polaroid instamatic camera.

Jeff has retained the original penciled checklists we used to tally a day's count (see Appendix Two). Dr. Severinghaus set the goal of reaching one hundred species in a day. Listing, keeping track of the numbers of different species, a "life list"—the seed was sown.

I was invited back the next spring. During that trip, I became hooked on birding. I vividly recall my first sighting of a Bald Eagle in Everglades National Park, as well as the wildly colorful birds like the male Painted Bunting and the Spot-breasted Oriole that nested in the Severinghauses' front yard. These special moments were my "spark birds." It was the ability to instantly recall the names of these birds the second time around that I found so fun and competitive. Yes, birding can be a competitive sport.

I'm frequently asked why birds interest me. Birds come in all shapes, sizes, and colors of course, and they sing. But most significantly for me—they can *fly*. Which means they can travel anywhere at any time. They are free and independent. And although I can't fly, birding does take me to interesting places, which is one of the best things about it. Birds also tell us a great deal about the health of our ecosystems. They respond rapidly to climate change, and bird distributions can be an early warning signal of changes that will impact human populations.

George (March 1968, Everglades National Park, Florida)

3. ABA Area

In 1968, at the age of ten, I set a goal to see 600 bird species in the United States by the impossibly old age of fifty. Why 600 bird species? Back then, there were a total of 645 identified birds in the United States and Canada. It seemed to me that spotting 600 bird species was akin to a Major League Baseball player hitting 500 career home runs. While Babe Ruth held the career record of 714 homers, only a handful of players had reached 500. So, seeing 600 bird species was achievable (like 500 home runs), and certainly *very* special.

The American Birding Association (ABA) was founded in 1969, only a year after I set my personal goal. It is a nonprofit organization dedicated to recreational birding in Canada and the United States. It has been called "the standard-bearer for serious birding in North America."[1]

Originally focused on finding, listing, and identifying rare birds, the ABA now seeks to serve all birders with a wide range of services and publications. It is the birding world's governing body. It recognizes all bird species seen in North America, which is defined as the land extending northward from the northern border of Mexico, plus adjacent islands and seas within 200 miles off the coast. (It excluded Hawaii until 2016). The ABA does not count sightings from Bermuda and Greenland—both places I have birded. I'm not sure why Mexico was left out.

Many people ask how many birds one can realistically see in the United States and Canada. It's a great question. Every species is assigned a number by the ABA between 1 and 6, based on its frequency of appearance. Code 6-birds are extinct, 5s have been recorded five or fewer times ever, and 4s constitute three or more records in the last thirty years. So, subtracting the 4—6s from the official ABA total leaves approximately 750. That is my answer.

Many birders share an obsessive-compulsive trait and, in general, like to keep lists, whether it is a list of birds seen in their yard or birds seen in a day, a year, or a lifetime. And it seems that another trait birders share is gloating about their United States/Canada (aka "ABA Area") life list. I suppose one's *world* list is the holy grail. There are approximately 10,600 species of birds in the world, and there are a couple of people who have seen more than 9,000. But most of us don't have the time or money to travel the world and bird, so we've settled on the ABA Area life list as one key measure of a person's true birding interest.

1 Scott Weidensaul, *Of a Feather: A Brief History of American Birding* (Harvest Books, 2008).

4. 1970s–1990s

George (July 1972, Oregon Coast)

George (May 1982, New Jersey)

In the summer of 1972, I attended the National Audubon Society Convention in Denver, alone. Flying to a gathering like that was not a typical summer activity for a fourteen-year-old, but I was hooked on birding, and my parents trusted that I could handle being the only person under twenty-one in attendance. The highlight of the five-day trip was birding with the legendary Roger Tory Peterson at several locations, including the Pawnee National Grasslands and Rocky Mountain National Park. I recall the group finding a dead Common Poorwill and learning that deceased birds do not "count" for one's life list. I was disappointed. It took me many years to find a living specimen to add to my list, as it is a secretive member of the nocturnal nightjar (*Caprimulgidae*) family.

My friends were supportive of this hobby, and I never felt odd or awkward about my interest in ornithology. I want to believe that my fascination with nature and birds rubbed off on my family and friends. Daughter Izzy once identified a Red-tailed Hawk on campus to the amusement of her schoolmates. My son-in-law Hunter Pawloff recently chased a Western Tanager in Manhattan. My friends Lindy Snider and Lee Warden periodically text me about birds. After questioning my "bird thing" while we were high school classmates, my friend Jay Pitocchelli earned his Ph.D in biology/ornithology and became a tenured professor and professional ornithologist.

I considered attending Cornell University to study ornithology. I also wanted to play Division 1 soccer, but Cornell was a top-ten NCAA team in 1974, and I felt I had no chance to play there. Jeff Dingle, an early employee of the Peregrine Fund (now known as the World Center for Birds of Prey) in Boise, Idaho, did attend Cornell.

I chose Princeton partly because five of my best friends (Jay Bennett, Chris Loughran, Henry Maguire, David Reed, and Jeff Shaw) were also matriculating there. I considered majoring in population biology, Princeton's equivalent to ornithology. It would have required taking organic chemistry with future doctors and chemical engineers at a ghastly early time three mornings per week. I decided to take a safer route to graduation by majoring in social psychology, headed to Wall Street (actually, 110 William Street), and kept birding as a lifetime interest.

My first job out of college was in New York City. Central Park and surrounding areas like Jamaica Bay are outstanding locations to search for birds, particularly migrants. I saw my first and only Tufted Duck in Central Park's reservoir, better known for Jackie Onassis sightings than for rare birds. Despite living in Manhattan, I was able to feed my hunger for birds in those early post-university years.

I will admit that after reading Jim Vardaman's 1980 book *Call Collect, Ask for Birdman*, in which he recounts his attempt to see 700 species of birds in America in one year (a "big year"), I was tempted to quit my new job and try a big year. But the cost seemed just too prohibitive both financially and socially.

In order to satiate my appetite, I spent many weekends visiting my parents and taking trips with my father to local spots in Pennsylvania, and the Jersey Shore. I recall dialing the Pennsylvania and New Jersey bird hotlines on Monday mornings and listening to Armas Hill of the Academy of Natural Sciences provide the past weekend's sightings.

George and Dr. Jay Pitocchelli (Ludlow, Vermont)

The problem was these were weekly tape recordings and by the time the next weekend rolled around the cited birds had likely moved on. Memorable sightings included my first Gyrfalcon at Stoltzfus Quarry near Lancaster, Pennsylvania, and my first Purple Sandpiper on a jetty next to Cape May's iconic lighthouse.

The day we saw that "lifer" (aka first) sandpiper, my father had taken both my younger brother, Wistar, and me on a day trip to the Pine Barrens to photograph orchids. We had time to add birding to the agenda, so we trekked to the shore. After scanning the coast for ducks, we spotted our Purple Sandpiper. A relatively common bird, but a lifer for me. As we headed back to the car, Cape May Bird Observatory's director, Pete Dunne, approached us with a band of eager birders in tow.

Pete was already a well-known author and ornithologist and someone I admired and hoped to know. As we passed, he politely asked: "Hi guys, seen anything good?" Before I could answer, my brother blurted: "Yes, my brother saw a King Eider just past the jetty." Pete's eyes widened, and the group immediately hustled to the water's edge to find what would have been a super rare bird. Wistar knew enough to be dangerous. I looked at him and then at my father. There wasn't a King Eider out there. My credibility was shot. I had to laugh and motioned my brother and father to the parking lot so Pete wouldn't associate me with the nonexistent duck.

George and daughter Sarah Wood

Son Cary Wood

Balancing career and family slowed down my quest, but I never lost track of my ABA Area life list. I retained my original handwritten checklists until the advent of computerized lists in the 1990s. Now Cornell University's Laboratory of Ornithology, through eBird, maintains a dynamic up-to-the-minute population census of birds around the world. At the same time, it keeps individuals' records so long as you subscribe to their app. It is simple and awesome.

I exposed my kids to the joys of nature and birds. Periodically, I took them on day trips.
I recall spontaneously loading Sarah, Cary, and Tyler into the minivan in a snowstorm to drive an hour north of our residence in Columbus, Ohio, to see my life Long-eared Owl in the Killdeer Plains Wildlife Area. I also added birding weekends to business trips. Conveniently, I had a client in Houston, so the trip to see my first Whooping Cranes in Aransas National Wildlife Refuge was simple.

During one of those business trips to Texas, I successfully chased a female Crimson-collared Grosbeak that had been reported on the rare bird alert airwaves. Within minutes of finding this bird on my own, I met a serious birder who had just flown in from California to see this specialty. His name was Jon Dunn. Yes, the now-infamous bird guide and coauthor of the National Geographic field guides. We exchanged business cards, and as a result, I took several WINGS weekend birding trips over the next few years to wonderful birding hot spots. Among those destinations were the Dry Tortugas, Florida, with a very young Michael O'Brien co-leading, and Southern Arizona, coastal California, and the Rio Grande Valley, Texas, with now-legendary birder-artist-author David Sibley, before he published his *Sibley Guide to Birds*; always adding to and tracking my life list.

By 1994, I had seen 500 species in North America. It required a pelagic trip off San Diego, but the sighting of a Black-vented Shearwater put me in a certain class of birding enthusiasts.

Six hundred species seemed years away. And it was. My fourth child, Karly, was born in 1995. I was now a stay-at-home dad with four kids under the age of eight. When would there be time to bird, and bird for relative rarities? The answer is—there wasn't time! But, as I stated earlier, there were many opportunities to introduce my own kids to the outdoors (although H-I-K-E became a four-letter word for one or two of them at times) and expose them to birding—and even falconry.

I'm not sure I detect a smile on any of the kids on this holiday card. The Harris's Hawks, endemic to Texas, are used to hunt rabbits.

Cary, Sarah, Karly, and Tyler Wood with Harris's Hawks (Gleneagles, Scotland)

5. Quest for Number 600

In 2010, I was playing basketball with Karly, then fourteen years old. I encouraged her to try her best, pursue her goals—the usual positive parenting remarks. Then she turned the tables on me. "Dad, whatever happened to that goal of yours? The one about seeing a bunch of birds." I told her that the goal was seeing 600 species by age fifty, and since I was now fifty-two, I had missed my opportunity. She looked at me and flatly stated, "Well, we lived in England for six years. Subtract that and you're only forty-six years old. I think you should pursue your goal." With that challenge—or, let's say, endorsement of my addiction—I reviewed my list, which was somewhere in the 560s, and embarked on a crusade to reach 600 as fast and responsibly as I could, given family duties and budget constraints.

In June 2011, I attended a birding festival in Carrington, North Dakota, and picked up five new birds among the prairies and potholes to attain 590 species. Funny, my middle name is Carrington. Was this kismet? The memorable moment of that trip was birding with the legendary Bill Thompson III, editor of *Bird Watcher's Digest*, and his accomplished wife, Julie Zickefoose. We set a "big day" record for me, recording 119 species in North Dakota of all places. My goal of 600 was within reach. A weekend trip off Half Moon Bay, California, in August brought me several new sea-going birds, and my number stood at 595.

The North American Rare Bird Alert (NARBA) network alerted me by e-mail to several autumn specialties. Four spontaneous trips to see these birds in Florida, Louisiana, New Jersey, and Massachusetts . . .

Mangrove Cuckoo *(Black Point Marina, Florida)*

American Flamingo *(Pass-a-Loutre, Louisiana)*

Barnacle Goose *(Newbury, Massachusetts)*

. . . increased my count to 599 in mid-November. Now my family and close friends demanded that number 600 be really special. But what bird should it be?

6. Shaw's Goose

***Pink-footed Goose**: number 600 (Windsor, Nova Scotia)*

A birding movie, *The Big Year*, based on Mark Obmascik's 2004 book, was released October 14, 2011. At the end of the movie, actor Steve Martin tells actor Jack Black that the very rare Pink-footed Goose has been spotted by his "friend Jeff Shaw." My twenty-one-year-old son, Cary, and I looked at each other and nearly jumped out of our seats. My great friend since kindergarten at Haverford is named *Jeff Shaw*. What are the chances? The answer to the question of what number 600 should be was obvious to the Shaws, who had seen the movie. Jeff's clever son Max told me that Number 600 had to be the Pink-footed Goose.

But this bird lives in Greenland and does not frequent Canada or the United States every year. While sitting with Jeff and his family at the Princeton-Yale football game, I googled "pink-footed goose, 2011, North America" and, amazingly, one of these birds had been frequenting a Nova Scotia farm with thousands of Canada Geese.

I could not leave town for five days, but on November 18, I flew to Halifax, where two friendly, knowledgeable birders helped me look for this wayward goose in very chilly weather. It was allegedly feeding on soybeans a mile from the airport. And I repeat . . . with *thousands* of Canada Geese. We divided into two cars half a mile apart with walkie-talkies. I was in one car with Dave Currie, and we'd been scouring the fields for maybe twenty minutes when the walkie-talkie crackled. John Robertson had our bird, and it was flying our way with ten or so Canada Geese. I was able to identify it with my binoculars, but, of course, I wanted a photo. I aimed my telephoto lens in the direction of the flock and clicked away.

Luckily, it flew overhead . . . and I captured it without leaving the passenger seat. Note the lack of a white face patch and no black in the neck, which are key field marks. The color of the legs is difficult to distinguish.

I love the birding community. Complete strangers to me a few hours earlier, John and Dave had the champagne ready! Class acts.

Jeff and Jeanie Shaw and George with the bird guide open to the Pink-footed Goose page.

Bird Party! (Bryn Mawr, Pennsylvania)

We celebrated the 600 milestone with a festive "bird party" at our home. Grey Goose vodka with a splash of pomegranate juice was liberally consumed as it was used for the signature drink of the evening—the Pink-footed Goose. I was showered with unique gifts from generous family and friends, including a pair of pink sneakers. Thank you Polly and Joe Knowles.

Mike Weilbacher wrote an article about my "600" achievement in our local newspaper, *The Main Line Times*. I was flattered, of course, but more importantly the article's final paragraph struck a chord with me. " Maybe he should shoot for 714 and become the Babe Ruth of local birders."[2]

There has been an increased interest in ornithology since Roger Tory Peterson's days in the 1970s. This is partly a result of improved technology producing affordable and easy-to-use digital cameras, spotting scopes, and binoculars. There is also the ongoing development of the ecotourism industry. Remote Alaskan islands were frequented by more birders and many new species were added to the official ABA Area list.

And with the advent of social media, one's phone now has field guide apps and rare bird alerts that can be programmed on an hourly basis. For me, 700 became the new 600. Yes, inflation, so to speak, devalued my original goal of 600, and now I had my sights set on 700. If I was not serious before, I was now—maybe even obsessed.

2 Mike Weilbacher, "Meet George Wood, the Mel Ott of birding," *The Main Line Times*, February 15, 2012

7. Winter in Washington

Andrew Bailey and author Brian Bell

I was obsessed but not to the point of sacrificing my job. I work for my alma mater, The Haverford School, in alumni relations and fundraising. My employment requires travel, and sometimes I can satisfy work responsibilities and pursue my hobby. In February 2012, we visited a favorite donor in the San Juan Islands, Washington. I persuaded my colleague Andrew Bailey to join me on the weekend to ride the ferries with a local bird expert, Brian Bell, and search for winter avians. We saw many black, white and gray seabirds that blended in quite well with the wintry sky and chilly water. Despite difficult conditions, we were able to record the diminutive Ancient Murrelet, lifer number 608.

8. English Plover

Natural disasters can have one benefit: they sometimes knock birds off course and bring surprises to the birding community. Hurricane Sandy delivered at least one mega-rarity to the East Coast from England—the Northern Lapwing. I was in the middle of a thirty-day Bikram Yoga challenge when I received the news of a lapwing on the end of Long Island. A dilemma of sorts, so I drove to Bridgehampton after my evening yoga class, slept a couple of hours and spent most of the next day searching for this Eurasian plover with many other eager birders. Nobody reported seeing it that day, and I returned to the yoga studio that evening and maintained my consecutive-day streak. But I wasn't happy.

Northern Lapwing *(Bridgewater, Massachusetts)*

Two weeks later, a lapwing was reported in a prison farm field in Bridgewater, Massachusetts. It was Thanksgiving Day, and as the dinner hosts, we had plenty of family duties. I didn't dare bring up the bird-chase topic until the next day. We were hosting pretty much the same crowd Friday evening, so I asked my wife, Nina, the inevitable question. To her credit, she said "Sure, after everyone has eaten the main course, just announce to everyone why you will miss dessert" or something to that effect. It was not exactly a surprise to my family, and they gave me their blessing. I told them that if everything worked out, I'd be home by early afternoon on Saturday —and I was.

9. Paragraph of Fame

I had to travel to the Sax-Zim Bog, north of Duluth, Minnesota, *in January*, to find my lifer Great Gray Owl. Okay, technically, this was my second sighting of a Great Gray Owl. I'd had a poor view of one in Idaho with Denver Holt several years earlier, but it was not satisfying. Owls are mysterious . . . like sharks. They are ubiquitous but rarely seen. The Great Gray Owl is a favorite of mine that I had admired for years but had seen only in photos and on television. It is larger than our much more common Great Horned Owl.

After seeing the owl from a distance with our group of Sax-Zim Boggers, (you see, it was an actual birding festival I attended), I went back to the lodge with the group to listen to the keynote speaker talk about woodpeckers. No offense to the topic or presenter, Paul Bannick, but I really wanted a good photo of this special bird.

So, I ditched the group, seemingly unnoticed, and drove back to the general area where we'd seen the Great Gray Owl earlier that day. I was not alone. Another fellow had the same idea. Gustave Axelson. We both spotted the bird at the same time and high-fived after taking a bunch of photos. It was 5:00 p.m. and the sun was setting.

This other naughty festival participant turned out to be a Cornell Lab of Ornithology researcher and part-time reporter for *The New York Times.* He was on assignment covering this unique bird gathering in ten-below temps in northern Minnesota. He was amused by my little escapade and my reaction to photographing the owl—so he wrote about me in *The New York Times.*

"We admired the gray concentric circles on its massive head, its fiery yellow eyes, and the white bow tie plumage below its chin. Another owl dived into the grasses and pounced on a rodent. A Discovery Channel live performance. 'Got it!' whispered Mr. Wood as he admired the images on his digital camera. 'Oh man, I'm going to print and enlarge these babies. They're going up on the wall at home.'"[3]

Great Gray Owl *(Sax-Zim Bog, Minnesota)*

3 Gustave Axelson, "In a Minnesota Bog, a Festival of Birds", *The New York Times*, January 20, 2013

10. Florida Specialties

The Florida Scrub-Jay is a threatened species endemic to Florida. It is the *only* bird that occurs just in Florida. It has lost over 90 percent of its 1970s population due to the development of its habitat (scrub-oak, sandy soil). Only a couple of thousand exist. In other words, this bird is special. Conveniently, I was in the Palm Beach area over Presidents' Day weekend. February 15, placed right between Valentine's Day and Nina's birthday, presented me an opening to chase this bird twenty minutes north in Juno Beach. Undaunted by the unusually heavy rain, I walked the fifty-acre nature spot with my iPhone in hand, playing the call of a male scrub-jay on my bird app. Two hours later, my list comprised exactly one Gray Catbird and a Little Blue Heron. Not the most productive morning, and rather an expensive one. Too consumed by the potential sighting of this eleven-inch rarity, I forgot that an iPhone does not tolerate rainstorms. It died before I reached my car.

Florida Scrub-Jay *(Jupiter, Florida)*

Not to be discouraged, I noticed my notes indicated that a few miles farther north, the Jupiter Ridge Natural Area was known to harbor a pair. The jay would be hard to miss as it looks and acts like our Blue Jay. Amazingly, the rain subsided, and an hour later I was wandering an even larger swath of nature, thankfully protected from the encroaching breeding populations of Hess gas stations and Denny's restaurants. With only thirty minutes to explore several hundred acres I headed to the spot that looked most like the textbook area for a Florida Scrub-Jay. Within three minutes I heard the diagnostic call and saw a speck of a bird on a distant telephone wire along Highway 1, about a mile away. My heart raced, and, incredibly, the bird alighted and flew toward me into the preserve. It landed ten feet from me. Where's my camera? I had left it in the car. I'll use my iPhone. No! It died, remember? But who would believe me without a photo? I hustled back to my car, slowed down by the sandy soil and the fact that I had not run very much lately. Five minutes seemed like an hour, but I returned to the spot to find two jays displaying typical behavior: one perched like a sentinel watching for predators while the other foraged in the sand for stored acorns.

Monday, February 18, was Presidents' Day, which meant I did not have to return to work quite yet. For two months I had been monitoring the North American Rare Bird Alerts and the presence of a rare female Western Spindalis frequenting the interior of the Key West Tropical Forest & Botanical Garden. That species used to be called a more familiar Stripe-headed Tanager. Why the birding authorities changed its name, I haven't bothered to google. But I do know that it isn't supposed to be in the United States. Home is Cuba or the Cayman Islands or somewhere far south. I had to see one. I contacted Larry Manfredi, a local hotshot birder from South Florida, and prepared for a long day of driving, hoping it would be capped by a stunning sight and photograph of this little gem. Late Sunday night I learned my guide, Larry had missed his flight out of the Bahamas, so I had to either scrap the trip (are you kidding?) or go solo.

The 3:30 a.m. alarm came quickly. To arrive at sunrise and have the best chance of seeing this bird, I needed three hours to make the drive, absolutely obeying the island speed limits the entire well-policed way. I arrived at the Key West Tropical Forest & Botanical Garden promptly at 6:45 a.m., only to learn that its gates did not open until 10:00 a.m. A lap around Fort Zachary Taylor, a brief peek into the bikers' world at the crowded Waffle House, and a bold use of the island's fanciest hotel's lobby for free wifi killed two hours. But I finally broke down and ordered a breakfast at a Denny's with tables that had not been wiped since Hurricane Andrew in 1992. That was hard to stomach, even more difficult to admit to you.

A Roseate Spoonbill sighting overhead put me in a better mood. This member of the Ibis family has a pink coloration from the food it consumes. Aquatic crustaceans like shrimp contain pigments (carotenoids) that cause a spoonbill's feather's to turn pink.

Roseate Spoonbill *(Key West, Florida)*

Ten o'clock finally arrived, and for the next three hours I searched in vain for a six-inch, non-vocalizing, gray-brown drab bird with a thin, high-pitched, infrequently uttered call. Females don't need to sing to attract attention—that is the male's burden. The gardens were "natural," which meant no paths and tons of tangles, and, even worse, it was popular with humans. Talking humans. I did not have a chance. No spindalis. I needed to make it to Palm Beach by evening, so I commenced the 200-mile return drive more than disappointed.

Florida City is an unsightly growth on the edge of the Keys and the Everglades. But it has plenty of rest stops where birds perch on wires and squawk at RVs.

Only recently, I had learned that the Common Myna, an invasive bird from Asia, had not only established a viable population in South Florida, but those birding authorities would let me count it on my life list.

"It is a large, stocky starling with a black hood and a patch of bright yellow bare skin behind the eye. Its bill and legs are yellow. Extremely vocal, Common Mynas mimic other birds and make a tremendous variety of gurgles, whistles, grating or liquid notes." [4]

A quick glance at the hordes of grackles, starlings, and pigeons produced two mynas. I couldn't believe it. Number 637. Not a bad day.

4 iBird Plus app

Common Myna *(Florida City, Florida)*

11. Texas Slam

Spring break afforded me the opportunity to bird the southernmost portion of Texas. Starting with Aransas National Wildlife Refuge on the Gulf Coast, the winter home of the 300 or so wild Whooping Cranes left on earth, we traveled down to the tip of Texas, literally the Brownsville City Landfill, and then up the Rio Grande a couple of hundred miles. A week of birding across 1,200 miles of Texas vastness, in a van with eight other bird enthusiasts, produced 203 different species, including four lifers for me. Not to mention some memorable human-interest stories that will have to be a topic for another time.

I did not see my number-one target bird for this trip, the one that tipped the balance in my favor when conducting a pre-registration "cost-benefit" analysis. Ferruginous Pygmy-Owl will continue to be a nemesis bird for me. I am 0-for-three lifetime looking for this fierce little daytime hunter.

Whooping Cranes *(Aransas National Wildlife Refuge, Texas)*

However, a second-day sighting of the rare Aplomado Falcon, followed by a Clay-colored Thrush at the Sabal Palm Sanctuary feeder the next day gave me the excuse to sample the local beer on the third evening and toast myself. None of the others imbibed, so I limited my intake to dos cervezas.

The tiny Northern Beardless-Tyrannulet has been one of my favorite names in the bird-guide since I can remember. Our keen leader (Chris Benesh of Field Guides) picked up its call at Santa Ana National Wildlife Refuge, and within minutes I had viewed but not photographed this small, innocuous, always-moving, grayish-olive flycatcher.

I must mention the skill of our leader. While viewing a collared peccary, also known as a javelina, in his scope, Chris somehow called out a distant, approaching-from-behind Gray Hawk. Not a lifer for me, but a real crowd-pleasing rarity and a reminder that I am merely an amateur birder. I never would have picked up this sighting, especially with my mind fixated on the strange mammal grazing in front of us.

The final new bird for me, an Audubon's Oriole, is a striking yellow and black bird found only in South Texas. It is not at all common, and it took some work playing recordings and listening for a whistle note to lure it in. Playing recordings is an accepted practice unless the bird one is seeking is endangered or in the middle of its breeding season. My new friend, "Dr. Jack," a young Thailand citizen currently on an infectious disease medical fellowship in Houston, whispered over to me and pointed upward. Bingo.

Audubon's Oriole *(Laredo, Texas)*

12. Expen$ive Godwit

For over three weeks, a vagrant Black-tailed Godwit had been seen every day at Chincoteague National Wildlife Refuge in Virginia. I simply could not remove this bird's image from my obsessive-compulsive-disordered (?) brain, so I signed out for a vacation day and took off at 3:00 a.m., knowing I absolutely had to attend a staff meeting at 2:00 p.m. I did write *Virginia*, didn't I? Two hundred miles one way. I traveled down the entire spine of our first state, passed through a portion of Maryland, I think, and hit the refuge at sunrise. I also made it back to Haverford for my meeting. That left very little time to search for this pretty large, conspicuous shorebird. (And please don't diss me about carbon footprint, etc. We are talking Black-tailed Godwit for crying out loud . . . and I drive a hybrid).

Oh, the bird. I didn't see it. Trying to make myself feel better, all day I kept thinking of the guy from Long Island who arrived at the same time as I did with his teenage son. He must not have slept at all the night before. Misery loves company. Of course, they may have spotted this rarity after I departed, once the tide let out and exposed the mudflat feeding ground. Maybe I should have checked the tide schedule before I left home.

Of course, I had to try for the Black-tailed Godwit again. It was seen Thursday—the day I "dipped" (missed it)—Friday, and Saturday. Sunday morning was wide open, so off I went. Unfortunately, only a mile from the Chincoteague Refuge entrance, I was abruptly stopped by Mr. Adams of the Accomack County police force. "Good morning, sir. Yes, I was probably driving too fast in anticipation of seeing the rare bird in your refuge." "You have a bird in the car? That is why I clocked you going forty-three in a twenty-five-mile-an-hour zone?" he responded with a perplexed look and an emphasis on the latter half of that sentence. He didn't get it.

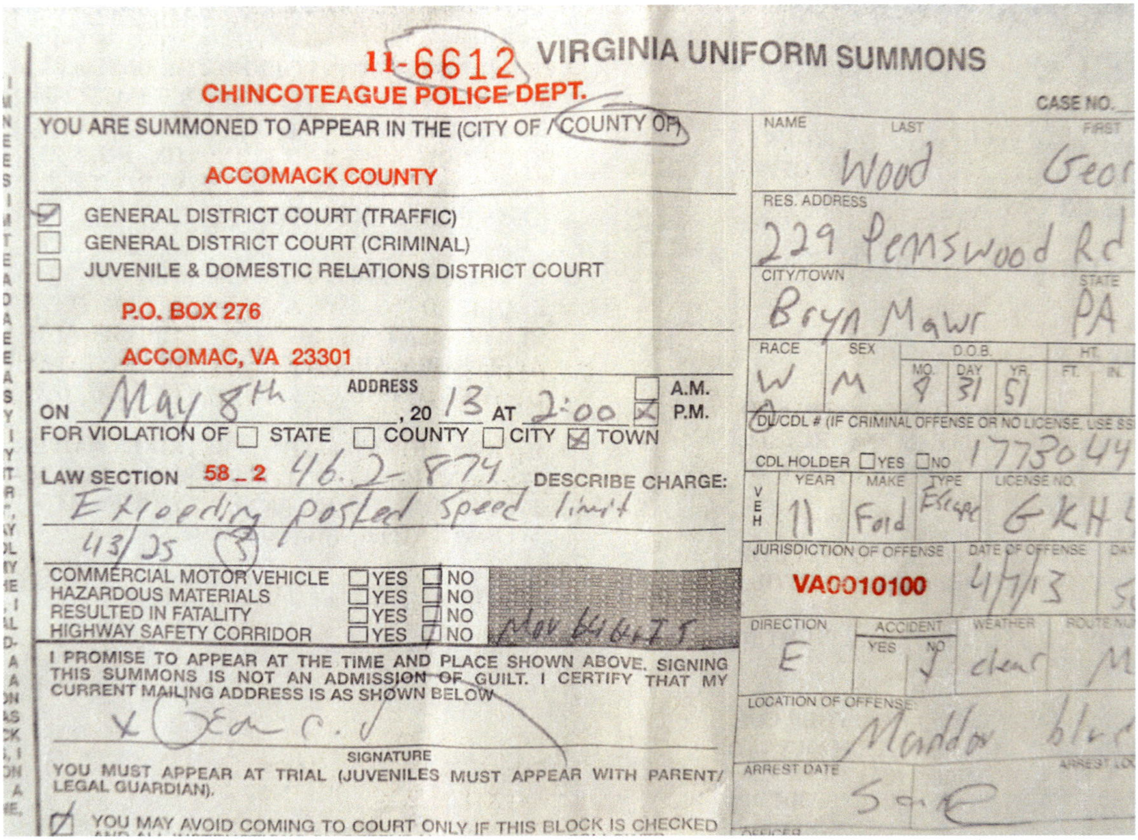

11-6612 VIRGINIA UNIFORM SUMMONS
CHINCOTEAGUE POLICE DEPT.
CASE NO.
YOU ARE SUMMONED TO APPEAR IN THE (CITY OF / COUNTY OF)
ACCOMACK COUNTY
☑ GENERAL DISTRICT COURT (TRAFFIC)
☐ GENERAL DISTRICT COURT (CRIMINAL)
☐ JUVENILE & DOMESTIC RELATIONS DISTRICT COURT
P.O. BOX 276
ACCOMAC, VA 23301
ADDRESS
ON May 8th, 20 13 AT 2:00 ☐ A.M. ☒ P.M.
FOR VIOLATION OF ☐ STATE ☐ COUNTY ☐ CITY ☒ TOWN
LAW SECTION 58-2 46.2-874 DESCRIBE CHARGE:
Exceeding posted speed limit
43/25
COMMERCIAL MOTOR VEHICLE ☐ YES ☐ NO
HAZARDOUS MATERIALS ☐ YES ☐ NO
RESULTED IN FATALITY ☐ YES ☐ NO
HIGHWAY SAFETY CORRIDOR ☐ YES ☑ NO
I PROMISE TO APPEAR AT THE TIME AND PLACE SHOWN ABOVE. SIGNING THIS SUMMONS IS NOT AN ADMISSION OF GUILT. I CERTIFY THAT MY CURRENT MAILING ADDRESS IS AS SHOWN BELOW
X Geo C. W
SIGNATURE
YOU MUST APPEAR AT TRIAL (JUVENILES MUST APPEAR WITH PARENT/ LEGAL GUARDIAN).
☑ YOU MAY AVOID COMING TO COURT ONLY IF THIS BLOCK IS CHECKED

NAME LAST FIRST
Wood Geo
RES. ADDRESS
229 Pennswood Rd
CITY/TOWN STATE
Bryn Mawr PA
RACE SEX D.O.B. HT
MO. DAY YR FT. IN.
W M 9 31 51
DL/CDL # (IF CRIMINAL OFFENSE OR NO LICENSE, USE SS
CDL HOLDER ☐ YES ☐ NO 1773044
VEH YEAR MAKE TYPE LICENSE NO.
11 Ford Escape GKH
JURISDICTION OF OFFENSE DATE OF OFFENSE DAY
VA0010100 4/7/13
DIRECTION ACCIDENT WEATHER ROUTE NU
YES NO
E clear M
LOCATION OF OFFENSE
Maddox blvd
ARREST DATE ARREST LOC
Same

Cost of bird chasing

Anyway, I shook off this minor setback and repressed the event. As soon as I reached the area where this shorebird had been seen recently, I noticed one pretty big bird about 400 yards out to my left, acting sandpiper-ish. I stopped the car and grabbed the binoculars. Looking through the car window, I knew it was a godwit for sure, but was the bill straight—as in the rare Black-tailed Godwit? Or slightly up-curved like its more common relative, the Marbled Godwit? I needed the scope. Moving quickly, I clumsily exited the car, grabbed the scope, and positioned it across the road closer to the bird. Yes, the bill appeared straight, but I wanted a photograph, and my camera was still in the car. As I hustled back toward the car, the godwit flapped and took off. No chance for a decent picture. I needed 100 percent confirmation for myself. Birders, like most golfers, live by an honor code, and the telltale, diagnostic marking would be the obvious presence of a black-and-white stripe on its flapping wings. My three-day-old Swarovski binoculars confirmed the Black-tailed Godwit. (Note: I am no longer "junior varsity," as someone recently labeled me because of my eight-year-old optics.)

13. Southeastern Arizona

Southeastern Arizona is the premier North American destination to search for owls and hummingbirds. The mesquite-lined washes in the seemingly endless desert around Tucson are dramatically framed by the distant Huachuca and Chiricahua Mountains to the south and east. These mountains, with some peaks exceeding 8,000 feet, contain riparian woodlands and pine-oak and spruce-fir forests, which translates into a huge variety of birds. I joined Dave Stejskal's Field Guides tour with hopes of adding a couple of lifers and photographing some amazing birds along the way. But the fact that I saw 160 species of birds and added one lifer (Plumbeous Vireo) in three days is *not* the subject of this writing. Rather, it is an admission of an amateur birder's embarrassing moment.

I was bringing up the rear of our nine-person group as we descended Miller Canyon deep in the Chiricahua Mountains. Dave was one hundred yards ahead of me and had already stopped and listened to the forest noises at the spot where I heard something interesting: a call that sounded like *pit-er-ick*. A quick search through the limited number of iTunes birdsongs I could recall came up with only tanager. It sounded like our Scarlet Tanager, but one has probably never been seen west of the Mississippi River. There are several western tanager species likely to be seen in this area, and we had already ticked the completely dark red Hepatic Tanager on our lists. Western Tanager, a common but brilliantly colored black, yellow, and red beauty, was on my mind. I hung back and searched the dense foliage for the source of this repeating call.

Since Dave had already moved ahead, and he pretty much identifies everything he sees and hears, I decided that whatever I found was not likely to be new for the group. Just then I spotted the orange head of the bird. Orange, like an oriole. I saw it for about three seconds, but, quite honestly, I hadn't done my homework and thought that it probably was just another one of the relatively common orioles we had already seen. Within five minutes or so, I caught up to the group, and Dave asked me what I had seen.

Flame-colored* x *Western Tanager hybrid *(Miller Canyon, Arizona)*

"Oh, it sounded like a tanager but looked like an oriole; it had an orange head," I casually replied. "An *orange* head? Are you sure?" he asked quite emphatically. "No question," I replied with confidence and a growing sense that maybe I had been too cavalier about this exchange. "The *Flame-colored Tanager* has an *orange* head!" Dave pronounced with bulging eyes. Uh-oh. The Flame-colored Tanager is a Mexican rarity seen maybe once a year in Arizona. Before I could move or respond, his six-foot four-inch frame marched back up the trail. After an unsuccessful attempt to re-locate my "oriole," we walked in silence back to the trailhead. I felt bad. I had missed a chance to identify something special and please our leader. I honestly would have traded several lifers for another crack at this bird.

Again, I hung back from the group. This time out of some embarrassment and shame; certainly not because I was interested in announcing a sighting the people in the front might have missed. I came around a bend and immediately knew the group was onto something good. I put my 10×42 Swarovski binoculars up to the sycamore tree where they were looking. Imagine my sense of relief and joy when I saw my "oriole."

Dave was the co-head of the Arizona Bird Records Committee. He believed this was a hybrid form, a rarity in itself—the Flame-colored X Western Tanager. Hybrids are not countable for the life list, but they are certainly worth a mention.

14. Nirvana

Looking from "sea-watch" over the Bering Sea to Siberia, Russia

Serious North American birders must witness spring migration in Alaska. Extreme listers will venture out to one of the islands with hopes of seeing North American birds that either migrate through or breed in the Arctic as well as rare Eurasian migrants that occasionally drop onto the island. I chose Gambell, the Siberian Yupik village on St. Lawrence Island in the Bering Sea, thirty-five miles off the Siberian coast. Gambell is a small, timeless Eskimo village where walrus, whales, seals, polar bears, and birds account for most of the native diet.

I was not disappointed, as I added a whopping twenty-one birds to my life list after five adrenaline-pumping sixteen-hour birding days on Gambell.

Gambell offers huge colonies of Parakeet, Least, and Crested Auklets (all lifers for me) on the cliffs. The point, or "sea-watch," provides a super-highway for passing seabirds—hundreds per minute—that fly very close to the island. But Gambell is most famous among birders for "the boneyards" and the rarities that are discovered in their craters.

The boneyards are the native dumps used for centuries to dispose of whale, walrus, and seal carcasses. The natives dig in the boneyards, looking for old ivory, which is quite valuable. Passing birds drop into the boneyards for both respite from difficult travels and nourishment from the bugs attracted to the decaying animals. Waterproof boots were mandatory!

Birding "the boneyards"

***Rustic Bunting** (Gambell, Alaska)*

A birder cannot visit Gambell on St. Lawrence Island without visiting Nome, too. Better known as the end point of the annual legendary Iditarod musher race from Anchorage, it is also a fantastic birding destination. Accessible only by boat and airplane, Nome is bordered on the west and south by the Bering Sea, and the north and east by miles of trackless wilderness. Unfortunately for me, winter still held Nome and its surrounding waterways in its icy grip. That translated into very few opportunities to search for Aleutian Terns, Spectacled Eiders, and other water-favoring specialties I wanted to add to my life list. However, a limited itinerary meant we had extra time to concentrate on the infamous eighty-five-mile-long Kougarok Road, which goes into the high tundra and mountains of the Seward Peninsula.

Kougarok Road is enchanting. It took us about one minute to sight lifer number 666, a roadside Willow Ptarmigan in a transitional plumage.

Willow Ptarmigan *(female molting from winter white to summer brown coat)*

By day's end I had added the rare Bristle-thighed Curlew and Rock Ptarmigan to my life list. If I maintained a mammal life list, I would have added another rarity: musk oxen. I spent my second and final day in Nome attempting to photograph the gorgeous but very elusive Bluethroat. It is a small thrush that breeds in northwestern Alaska and Eurasia and, for me, is a signature bird for the high Arctic.

I desperately wanted a photo good enough to print, frame and hang in our already overrepresented aviary of a living room. After hours of trudging through marshy willow scrub bogs, I captured one relatively stationary male.

Bluethroat *(specialty of the high Arctic)*

15. The Curllsville Strips

For me to reach the "700 Club" of North American species sightings, I knew I needed to have the Henslow's Sparrow. Each new bird from this point on was going to be a struggle—not to mention potentially expensive. The Henslow's is a breeder in the western part of my home state, and I have no good explanation for why I had never tried to see it during the four years we lived in Columbus and frequently drove to the Philadelphia area. An uncommon bird with a declining population, the easternmost part of its breeding range is 325 miles west of Haverford in the Curllsville Strips, north of Pittsburgh. It might as well be in Ohio. The Curllsville Strips site is an aging reclaimed strip mine grassland, privately owned but open to the birding public. Armed with a reliable report of multiple sightings that week, I could not resist the temptation to go out there early one morning. Music from the 1970s, intermittently peppered with ESPN sports radio, entertained me along the rather long stretch of I-80.

Within seconds of entering the strips site, I was greeted by a male Dickcissel, no shabby bird itself (I may have seen three in my lifetime), singing on territory. One minute later, as I drove slowly down the dirt road with the windows open, I heard the distinctive *se-lick* song of the Henslow's Sparrow in stereo. No, it was more like surround sound. Four or five of them were audible but hidden in the overgrown fields.

Henslow's Sparrow *(Curllsville Strips, Pennsylvania)*

And then good fortune struck. One brave male perched ten feet from my car. I was gone only twelve hours, having driven—do I dare admit it?—650 miles (sorry, carbon counters). It was a selfish act on my part. But on this day when my dad would have turned ninety, I shamelessly claim lifer number 669 in his honor.

16. Eight Lifers, Chiggers and Bear, Oh My!

Three days in the Santa Rita Mountains south of Tucson, Arizona, produced eight lifers. Without the enthusiastic, expert guidance of local birding professional Brian Gibbons, my trip would have been a bust.

Brian and I hiked miles and miles of canyon trails at elevations greater than 5,000 feet in pursuit of rarities that frequent North America only by way of southern Arizona, in the summer.

These specialties are basically Mexican birds.

Sulphur-bellied Flycatcher
(Madera Canyon, Arizona)

Black-capped Gnatcatcher
(Montosa Canyon, Arizona)

White-eared Hummingbird
(Miller Canyon, Arizona)

We were not the only birders on the trail up Miller Canyon in search of the White-eared Hummingbird. About a mile up the very steep slope I recognized eighty-two-years-young Sandy Komito, the person who held the big year record at that time; you know, the character portrayed by Owen Wilson in the not-so-famous 2011 movie of the same name. We talked briefly, and I gave him my business card. No, he never chose to correspond with me.

As I waited in the Phoenix airport for my red-eye flight to Philly, I was distracted by the itchy welts around my waist caused by chiggers. I have no idea how I hosted these microbes, but I blame the bear we encountered the day before. His sudden appearance on our hike up Madera Canyon dropped me in my tracks and left me sitting in some strange vegetation as I attempted to take a memorable photograph.

A brown black bear in Madera Canyon

17. SF Friends

Nina lived in San Francisco for twenty years, so a one-week summer trip to the Bay Area served many purposes. Rekindling friendships and making new friends, eating healthy foods like kale and soybeans, drinking healthy beers like Blue Moon, promoting Haverford, of course; did I mention birding? Basically, I had my binoculars around my neck the entire visit. It did not seem to bother my Haverford alumni—not even the conservatively dressed lawyers. My tolerant and generous hosts, Megan and Mark Horton, David Simpson, and Anne and Michael Parish, may have even caught a bit of the birding fever I intentionally tried to pass on to them. By the time I was finished talking (at them?), they were locating Great Horned Owls, distinguishing whimbrels from godwits, and even cutting out newspaper articles about birds.

Walks through Golden Gate Park, hikes up Mount Tam (twice) and exploring Point Reyes National Park produced many bird encounters but no lifers. A twelve-hour day on the infamous Debra Shearwater's Half Moon Bay pelagic voyage provided incredible cetacean sightings (blue, fin, and humpback whales), but alas, the rare Cook's Petrel must have forgotten that it was supposed to leave New Zealand by now.

I spent most of my time at the stern of the boat befriending Ross, the seventeen-year-old "chum guy." His job was to systematically toss popcorn and drip smelly fish oil over the back of the boat in order to attract the birds. Fortunately, no seasick passengers interrupted our conversations about seeing condors or chasing the Rufous-necked Wood-Rail. Talking birds with Ross, a young birder extraordinaire headed to the Cornell Lab of Ornithology the next month for a special young birders symposium, was as enjoyable as watching Black-footed Albatrosses glide alongside our converted salmon-fishing vessel. (The multipurpose boat was also specially licensed to dispose of cremated human remains.)

Black-footed Albatross *(Half Moon Bay, California)*

The Philadelphia Phillies and San Francisco Giants baseball teams were in a July slump with little prospect for improvement. My mini-slump would soon be snapped, though. A Hatteras, North Carolina, two-day pelagic trip—call it a double-header—was only two weeks away.

18. Pelagic Magic

In order to reach the goal of 700 North American species, I knew that I had to endure one or two of the legendary twelve-hour pelagic trips off Hatteras, North Carolina.

Trindade Petrel *(Hatteras, North Carolina)*

We were guided by famed skipper Brian Patteson, with the goal of seeing the rare Black-capped Petrel and Band-rumped Storm-Petrel. These birds are rarely seen from shore and breed in Cuba, the Azores, and the Canary Islands. Thirty miles off the port of Hatteras on the Outer Banks of North Carolina, still in ABA territory, the continental shelf drops off and meets the Gulf Stream. This combination of deep and warm water creates a dynamic ecosystem teeming with fish and birds. Brian took out our group from the Pennsylvania Society for Ornithology on two consecutive days. Eight-foot swells and thunderstorms did not deter him. As a matter of fact, it increased our chances for not only the two target birds, but also a rarity from the south. Black-capped Petrels were in abundance, but the little Band-rumped Storm-Petrel made only a couple of fleeting appearances. However, our sighting of the Herald (now Trindade) Petrel, a breeder on islands off Brazil, caused quite a stir on the boat. Yes, even the five or six seasick birders scrambled off their berths and stumbled to the stern. This bird immediately made the North American Rare Bird Alerts.

Note: when driving from Philadelphia, consult a map—don't rely on GPS. The trip took eleven hours (659 miles, a new personal one-day driving record not to be broadcast beyond you) as the GPS led me to the Ocracoke ferry. Since I did not have a reservation, I needed to double back, adding an extra one hundred miles to the trip. The GPS considers ferry crossings as pavement and thus uses it in its calculations for "fastest" mode. The ferry ticket taker is still laughing at my question: "Where is the bridge?"

19. Salton Sea Success

It had been a couple of months since I chased a rarity, as work took priority. However, the opportunity to experience the very unusual invasion of Blue-footed Boobies in California swept me away to San Diego. These birds rarely stray north into the Lower 48, but the sight of them as far up the California coast as Point Reyes continued to send shock waves throughout the birding world.

My birding guide from August's Arizona adventure, Brian Gibbons, drove from Tucson to San Diego and chauffeured me around San Diego and Imperial County to search for the boobies at the Salton Sea, as well as other potential lifers for me. The Salton Sea is a birder's mecca if not anything else. Formed in 1935 by the combination of below-sea-level terrain and an overflowing Colorado River, it is a ridiculously special migrant trap for thirsty birds. Two juvenile boobies planted on Obsidian Butte like ornaments on a Christmas tree proved a relatively simple find.

Blue-footed Boobies *(Salton Sea, California)*

The Salton Sea is famous among birders for another very special bird, the Yellow-footed Gull. It can be found in North America only in the Salton Sea. Its population is in decline, and we spent hours searching for one after ticking the boobies. Despite recent eBird reports and the coordinated efforts of a father-son duo from Boston, who clearly place birds above baseball, we struck out on our first afternoon.

At sunset, with no more birds to log nor a dinner/sleep plan, we drove up to a parked pickup truck with a California license plate that read "SWbirds." Its occupant didn't profile "birdwatcher," but this Stetson-topped, cowboy-looking guy happened to be surveying the field full of Sandhill Cranes with his very impressive binoculars. Not only did he recommend a wonderful fairly local place to spend the night (there aren't any towns near the Salton Sea), but he shared a precise location to likely find the one adult Yellow-footed Gull in the area.

Our dawn arrival the next morning produced this heart-pumping sight.

Yellow-footed Gull *(Salton Sea, California)*

The rest of the weekend was hardly a letdown as we pulled one Red-throated Pipit from a Tijuana River Valley sod farm full of American Pipits. In addition, we scooped up the Tricolored Blackbird and the recently countable African native Scaly-breasted Munia to bring my ABA Area lifetime count to 685. We spent the entire final day searching in vain for the Mountain Quail and Bell's Sparrow. Number seven hundred was in view.

20. NASCAR?

"Are you in town for the NASCAR championships?" asked the Miami rental car person. Strange, the people next to me on the airplane had asked me the same question. While I may consider my weekend birding-travel attire as neo–Marlin Perkins, (remember Wild Kingdom?), clearly my Haverford logo-laden look—with black "H" cap, black "H" vest, and black "H" carry-on—overshadowed my khaki shirt and pants combo and gave people the impression I was a walking "H auto-parts" sponsor rather than my intended conjuring of images of Steve Irwin, Jeff Corwin, or some other Animal Planet hero.

I traveled to Miami for thirty-six hours solely to add three non-native but countable resident birds to my life list. I added only one, the Purple Swamphen. But I traded in the opportunity to add two parakeets to my list in favor of a needle-in-the-haystack mission.

Purple Swamphen *(Miami, Florida)*

The adrenaline rush of potentially finding a Code-5 mega-rarity—a bird not reported publicly for various reasons, too confusing to report here, but known to my well-informed guide—is at least equal to that of any athletic contest I have ever played. At this point I did not have any Code-5 birds on my life list, and I had only a handful of Code-4s. For the birding aficionados: Pink-footed Goose, Barnacle Goose, Crimson-collared Grosbeak, and Northern Lapwing. As I stated earlier, the scale goes only to 5; Code-6 birds are extinct.

We spent all Saturday afternoon and Sunday morning searching for a foot-long, chunky, ground-dwelling bird that was very likely within a football field–sized area of jungle cut with several pathways. We circled and zigzagged for hours, and at one point we even found a very recently molted feather that may have belonged to this top-secret bird. Alas, we didn't find the Ruddy Quail-Dove.

I flew home to Philadelphia far from disappointed. Despite recording only one lifer and greatly exceeding my budgeted cost-per-new-species figure, I was grateful to have had a chance at being a part of birding history. Maybe next time.

21. Countdown to Number 700

My daughter Chloe suggested I post my sightings on the sunroom wall. Because it was brick, it was useless for photos and a perfect place for me to keep track and stay motivated to reach 700. We sold the house before I reached 700, but the current owners have kept my artwork.

22. December Delight

The Fork-tailed Flycatcher was supposed to be in Argentina. Southern Mexico is the northern edge of its normal range. However, once in a while this spectacular flier, whose forked tail is almost a foot long, wanders off course—3,000 miles! Ornithologists theorize that some individuals' internal clocks are spun around 180 degrees and thus they migrate north, as opposed to south.

The rare bird alert reported an adult male Fork-tail, considered a Code-3 by the ABA, frequenting the Connecticut River's Hadlyme ferry slip. Content with devouring plentiful berries and lingering insects, it had been hanging around all week. This was Connecticut's second Forky in three years.

I had planned on chasing the Code-5 Amazon Kingfisher in Harlingen, Texas, that weekend. But a lack of reliable sightings that week combined with ridiculously cold and wet weather forecast for southern Texas produced a moment of sanity. I postponed Texas and headed to Connecticut.

That day was the one-year anniversary of my birding blog, a concept hatched by my daughter Sarah, who encouraged me to document the final stretch of the pursuit to see 700 North American (ABA) bird species in my lifetime. I believe it was a polite way to inform me that she and most others would prefer to read about my quest on their own time, rather than listen to my spontaneous and frequent bird tales—or not pay attention to them at all.

Fork-tailed Flycatcher *(Lyme, Connecticut)*

I completely understood. So, she introduced me to the concept of a blog. With her guidance, I created birdtalesblog.com, and after that I tried to document my travels with humorous and informative wording always accompanied by at least one photograph.

Whatever the case, I am flattered friends and family read and seemed to enjoy the posts, particularly the photos. I honestly thought the earliest I would reach 700 would be in 2017, but 2013 turned out to be a really big year. As a matter of fact, the Cornell Lab of Ornithology reports that I saw 541 species in 2013 alone, ranking me 119th among those who submitted observations that year. I sent Cornell's eBird website all eighty-one of my checklists, which ranked me only 5,993rd, so many people do this birding thing seriously.

My mother was concerned that I would experience a mental let-down when I reached 700. Admittedly, I was obsessed about reaching that number, but I promised that once I did, I would slow down and concentrate more on good photographs as opposed to good birds. I wouldn't be bored; no worries. I was thankful she cared.

How many people have reached the 700 mark? My educated guess is about 1,000. They are generally professional guides, retired, or people like me—fortunate enough to have the resources and support from home to pursue a passion.

I want to emphasize that although bird listing is just a game, it is a wonderful excuse to enjoy and notice the incredibly interesting natural world that surrounds us. Of course, it is a good exercise in planning, execution, and persistence. Are any of my children reading this? The count stood at 693, and you bet I had a plan to hit 700 by Memorial Day.

23. The Great White Owl

Sometimes it's better to be lucky than good. A Snowy Owl rarely visits Pennsylvania. It is an Arctic bird that sporadically travels south of Canada. You may have heard or seen that a Snowy Owl was located in Central Park, New York, at the end of January 2021. The first time it had been seen there in 130 years!

When I received an alert on my phone that a Snowy was only forty-five minutes away from my residence in Pennsylvania, I dropped my packages and raced home to grab my camera and head west toward Spring City. You see, it was a couple of days before Christmas, and I was in the middle of shopping in Suburban Square, Ardmore. I'm sure the salesclerk was a bit miffed that I left him in mid-sentence, but I don't regret my decision. When I reached the designated owl spot, I could see, perched on a hill, a dozen tripods and scopes and cameras with hunched-over humans attached to the eyepieces.

I approached them quietly and watched them watch the owl about a half mile away. The owl was sitting on a recreational vehicle in a lot full of RVs protected by a barbed-wire fence. I asked a couple of the folks why they weren't closer to the bird. I really wanted a photograph, and I did not have a fancy zoom lens. The people remarked that the owners were tough, not nice, didn't like birders, etc. When I inquired if any of them had actually spoken to the owners, they replied no.

So, with their blessing, I walked down the hill and knocked on the office door. I was greeted by several employees who eagerly asked me what the fuss was on the hill. What were these people

Snowy Owl *(Spring City, Pennsylvania)*

looking at? I explained the situation and even showed the apparent owner a Snowy Owl on his computer. He enthusiastically stated that we needed to have a closer look at this thing. I followed him out the door, weaving in and around the multitude of RVs, until he stopped at one and ascended the ladder. He came face-to-face with our bird. After what seemed like minutes but must have been seconds, I politely tapped him on the back of his leg and asked if I could take his place and take a photo. He graciously agreed. No telephoto lens needed. This owl did not even wake up.

Of course, I do not always bird alone or with strangers on a tour. Being with friends while birding is a life highlight. High school buddies and college roommates have been force-fed birding stories and excursions for decades.

George with college roommates:
John Wallace, Henry Maguire, and Haden Gerrish (Vinylhaven, Maine)

24. Good Day Nanday

A weekend visit to see our friends Deborah and Dennis Glass in Palm Beach produced unexpected lifer number 694. Do you like how I casually dropped "Palm Beach"? I could get used to this. The Nanday Parakeet isn't super rare in Florida. It is an established South American escapee. The surprise was that I only recently learned that the ABA added this spectacular parakeet to the countable list.

Now, it wasn't a simple endeavor. I needed permission from my hosts and Nina to disappear only a few hours after I'd arrived. Truth is, they know me well enough and urged me out the door. Then I needed to search eBird and query if this bird had been seen and reported near there recently. The last post was more than a week old, and it was thirty miles away. Not terrible odds. But not good enough to bother bringing my digital camera. I left before dawn, and five hours later, after much driving with the windows down the better to hear these squawkers before seeing them, I was ready to return home and bask poolside in the warm weather.

But my sixth sense told me to try the Loxahatchee National Wildlife Refuge one more time. Three hundred yards before the entrance I noticed two large birds on the telephone wires just ahead. My heart jumped, and I steered my rental into a ditch, barely remembering to throw the gear into park; it wouldn't have mattered anyway, as I was nosed up against a hill and quickly snapped one photo with my iPhone before they flew off. I made it home in time for lunch and a celebratory beer. Life is good.

I went to Florida again with a chance to see four new birds. Unfortunately, three of them did not cooperate. The White-cheeked Pintail was not in the pond near Vero in America's first national wildlife refuge. The Black Rail was not calling in the "can't miss" marsh at 4:00 a.m. on either day despite reports that someone had seen it the day before and the day after I was there. And the Budgerigar was not at the feeder where a Norwegian vacationer had photographed it exactly ten minutes before I arrived. However, the White-winged Parakeets were in the date palm trees near Miami's airport, making up for the fact that I had missed them in that same spot last November.

The White-winged Parakeet is a native of South America. In the 1960s it was introduced to California and Florida, where feral populations became established from released and escaped birds. Hey, the birding authorities count it, so I did, too. Number 695.

White-winged Parakeet *(Miami, Florida)*

25. March Madness

While I was in Florida, the rare bird alert texts were flooding my iPhone. New Jersey's first-ever Eurasian Tree Sparrow was frequenting a feeder in Cape May. Fifty-seven people had reported it by noon on Sunday. Admittedly, this bird was never on my "need to have" list.

It was introduced to St. Louis, Missouri, in 1870 from Europe and has pretty much stayed as a local specialty there. It looks very much like our typical House Sparrow (an "LBJ," or little brown job) but with a rusty brown cap as the distinguishing feature as compared to the abundant House Sparrow's gray cap. Somebody was really paying attention to pick out this little fellow.

A business trip to Boston on Monday and Tuesday kept me 500 miles away from lifer number 696. On Wednesday, we finally closed on our chic new urban townhouse in Philadelphia. With a mind more focused on this weekend's adventure to the King Ranch in Texas, I took off for Cape May before dawn—despite the fact that no reports of our Midwestern visitor had been posted on the listserv since Monday evening.

Eurasian Tree Sparrow *(Cape May, New Jersey)*

I arrived at Cape May Point at 7:00 a.m. expecting to see throngs of birders. I was alone. "Am I the only birding adult benefiting from spring break," I wondered out loud. Maybe the birding authorities had already pronounced this wayward sparrow as an escaped exotic but failed to inform me. More likely, anyone who cared had already seen this very cooperative bird. I drove slowly down Harvard Avenue in search of number 113. "For Sale," "For Rent," barren, bird-wise until I saw one yard decorated with hanging, stocked feeders. It took me no more than twenty seconds to pick out this surprisingly beautiful avian from the crowd of Starlings, Mourning Doves, and House Sparrows.

While many were debating who would make the NCAA Final Four hoops tourney, I had placed a wager on my final four. Tip-off was the next morning in Harlingen.

26. Nemesis Bird

The only location in North America where one has a decent chance of hearing and possibly seeing the fierce Ferruginous Pygmy-Owl, a diminutive (six inches), diurnal hunter of small mammals and even larger southern invertebrates, is on the privately owned King Ranch at the bottom of Texas. The ranch is physically larger than Rhode Island, and despite the fact that naturalists have staked out several nesting areas, success is never guaranteed.

I left Philadelphia on a Friday evening and landed in Houston at 9:45 p.m., missing my connection to my destination, Harlingen. I absolutely had to be at the King Ranch's Norias Division gates at 6:00 a.m. Saturday or I would miss the special van tour I had reserved months ago. Thankfully, I did catch the delayed 10:30 p.m. flight to Brownsville. Brownsville is south of Harlingen and a two-hour drive to my morning meeting spot.

Of course, this revised itinerary assumed I could rent a car in Brownsville. There was not a single vehicle available to rent. While one side of my brain was rationalizing the expense of a two-hour taxi ride, my emotional half took over and blurted out: "I am a professional wildlife photographer, and I'll lose my contract with the King Ranch if I fail to show up." The sympathetic agent eyed the two carry-on bags weighing me down and promptly released a vehicle from the secret stash.

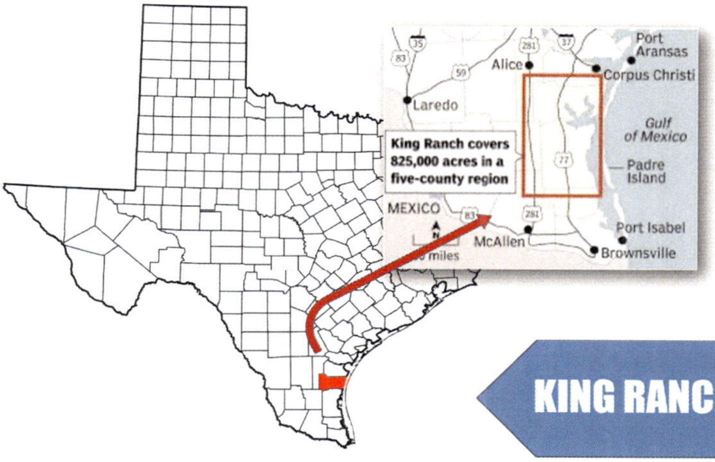

KING RANCH, TEXAS

Our van tour was relatively uneventful until our guide announced that we were in the Ferruginous Pygmy-Owl's territory. The group quietly exited the vehicle and scattered around the live oak and honey mesquite trees. Almost immediately, our guide announced he heard the owl's call, seemingly fifty yards away. It took several minutes but soon everyone confirmed they'd heard it. Surprisingly, the group began climbing back into the van even though we had not seen the bird. I pleaded with the leader to give me ten minutes to continue my search. Perseverance mixed with luck produced this ventriloquist sitting on a branch above my head.

I have really wanted to see this incredibly difficult-to-find little owl since the early 1990s. Many trips to Arizona and Texas produced scores of great birds and memories, but this elusive powerhouse stood out as the only North American owl missing from my life list. Number 697.

Ferruginous Pygmy-Owl *(King Ranch, Texas)*

27. Susquehanna River Delivers

Little Gull *(Wrightsville, Pennsylvania)*

The Susquehanna River had been hosting several unusual Code-2 Little Gulls, but I'll admit that I am not super confident in my gull identifications. Besides, the Susquehanna isn't exactly a creek. The fact that thousands of other similar-looking gulls—specifically Bonaparte's Gulls—were frequenting the same stretch of the river near Lancaster had discouraged me from even trying to locate our smallest North American gull for the previous two months.

However, three consecutive daily rare bird alert texts informing birders of a Little Gull hanging out between the bridges in Wrightstown motivated me to try. I banked some credit in the morning by vacuuming and mopping all the floors around the myriad of boxes cluttering our five-day-old residence in Philadelphia before suggesting my solo journey to the Susquehanna River. It worked.

As soon as I reached the John Wright Restaurant parking area on the shore between the two bridges, I set up my new Swarovski scope within speaking distance of a young fellow intently staring through a much more impressive optic setup. I spent ten minutes scanning the vast river, trying to find the one white bird with a black underwing. Nothing. I was experiencing a competitive birder's dilemma: keep searching on my own or ask the guy next to me for help. I chose the latter.

"Sure, I'm on it now, straight out about half a mile," he stated matter-of-factly. I thanked him and fumbled for my camera, which was twisted around my binoculars—both dangling and clanging against my tripod. The bird floated downstream with hundreds of its identical cousins until a certain point, where it flew back in the reverse direction. Its rounded gray wings, edged in white, contrasted brightly with the opposite black underwing. There was no doubt. . . number 698.

28. Secretive Quail

I am often asked if you can count a bird if you only hear it. Technically, the answer is yes, but my answer has always been no. Birders will debate the pros and cons, but for me it is clear—seeing is believing. I don't count "heard-only" birds. The Mountain Quail is one of those birds that is heard sometimes and seen infrequently. I have attempted to locate this secretive western quail several times to no avail.

Accompanied by an experienced guide, Rich Cimino, I traveled eighty miles north of San Francisco to prime Mountain Quail habitat. A steep canyon at 2,000 feet elevation, lined with a mixture of oaks and conifers interspersed among chaparral grasses, presented the perfect spot to hear a male Mountain Quail call from his customary sentry position: a rocky outcrop or tree snag. Within seconds of Rich's order to stop the car, we could hear the unmistakable *quee-ark* crowing call quite easily. But where was the bird? Thirty minutes elapsed, and we still had not located the persistently calling quail. Desperate to see this guy, I began climbing up the hillside

Mountain Quail *(Healdsburg, California)*

on all fours, much to Rich's amusement. Of course, by the time I had reached its presumed location, the calling came from an entirely different direction. So frustrating.

Probably holding back laughter, Rich politely suggested that we drive up the canyon. Sometimes these elusive creatures will forage along the shady side of the road. Admittedly, I thought we had missed our chance of seeing number 699. I started driving up the very windy road, my mind still on the hillside, envisioning what should have been. Suddenly, a dark hen-like figure flashed in front of me and darted around the corner. If Rich hadn't put the car in park after my quick exit, well, I don't want to contemplate the consequences. Anyhow, I hoped for the best and in seconds was rewarded. Number 699.

My phone is set to receive daily alerts from the NARBA birding gods when any Code-3, -4, or -5 is seen in North America. I had seen all Code-1s and -2s except a handful—Yellow Rail and Black Swift to name two—as of 2022.

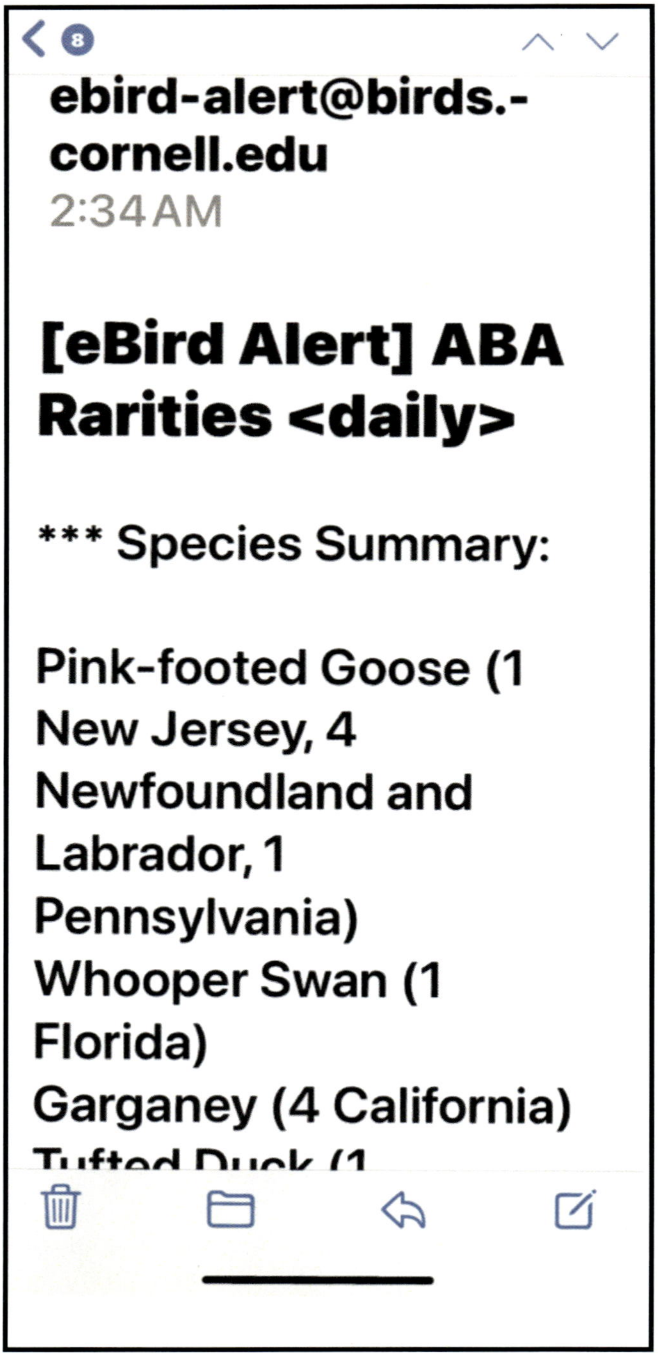
ebird-alert@birds.-cornell.edu
2:34 AM

[eBird Alert] ABA Rarities <daily>

*** Species Summary:

Pink-footed Goose (1 New Jersey, 4 Newfoundland and Labrador, 1 Pennsylvania)
Whooper Swan (1 Florida)
Garganey (4 California)
Tufted Duck (1

29. Pursuing the Black Rail

699½. What does that mean? The short version is that I heard the Black Rail but I did not see it. The longer version follows.

Nailing the Mountain Quail early Saturday morning meant that I had two days and nights to devote to the pursuit of number 700 before my Monday flight to Anchorage. At that moment, the Black Rail was the only bird within an eight-hour drive of San Francisco that was missing from my life list. There was a rare Red-billed Tropicbird flying around the ocean off San Diego—too far. Black Rail is a most worthy candidate for owning the coveted number 700 spot because this mysterious six-inch, super-secretive marsh bird is generally recognized to be one of, if not the most difficult birds to see in the wild. It rarely flies, as it prefers to creep around the pickleweed like a mouse. It does vocalize but almost exclusively at night. Of course, besides just wanting to see this bird for myself, I also felt added pressure to take a recognizable photograph to satisfy readers and cap off my wonderful blogging experience.

The bird breeds in the Corte Madera marshes within view of San Quentin prison. No, I didn't add jailbird to my list. Daylight birding produced some interesting sightings, like White-tailed Kite, Marbled Godwit, American Avocet, and a manta ray that rode the swift incoming tide a bit too long. But not a hint of any rail, not even the common Clapper Rail. Nocturnal birding is more productive when binoculars are supplemented with a high-powered light. Add the fact that the weather was calm and the full moon produced a very high tide that pushed the birds into a tighter space, and my chances were perfect.

Well, in fact, I heard several Black Rails calling over the course of many hours both Saturday and Sunday nights. At times they sounded like they were right in front of me. The noise they make is like no other rail or bird. *Kic-kic-kerr* repeated over and over is imprinted on my brain and may bother me for a very long time. The males have clearly perfected a strategy to avoid predators (herons, hawks, bobcats, birders?) while simultaneously calling attention to themselves for mating purposes.

I am 100 percent confident there were no other birders in the field using recorded calls, so Black Rails were, indeed, present. But I never saw one and thus remained at 699; or was it 699½?

30. Thanks to Toby

You are not likely to run into a Code-2 Aleutian Tern unless you are in the Aleutian Islands or at Headquarters Lake in the Kenai National Wildlife Refuge, 150 miles south of Anchorage. I received a tip that these maritime feeders had recently taken up residence on a small inland lake—a presumed lay-up. Since I had a free day before my Pribilof Islands adventure, I drove down the scenic Kenai Highway filled with optimism.

Upon arriving at the tiny observation platform, I was more than disappointed to see a huge lake void of avians. Where were the terns? I could see distant birds flying over what turned out to be a floating bog-island, but the smoke from a growing wildfire was beginning to erase the blue sky and cast a large shadow.

I hustled back to the refuge headquarters and convinced a very friendly refuge field person (Toby) to bring his 60x magnification scope to the viewing area. We were able to positively identify a couple of Aleutian Terns, but what about photos? I couldn't write my blog without proof of number 700. Toby discouraged any thoughts of using a yet-to-be-located form of water transport. 1) It was illegal for visitors, and 2) the lake was off-limits even for refuge personnel, as it was needed as a landing area for the increasing number of firefighting seaplanes. With that directive, he headed back toward his office.

As I paced the shore wishing for an impossible Aleutian Tern flyby, I noticed an upside down canoe in an enclosure. I hesitated but quickly made the right decision for me and somehow scaled the ten-foot fence without busting my optic equipment or my leg. The canoe's midriff contained two paddles and a life preserver. At least I would appear legal in someone's eyes if I were run over by a seaplane or swamped by the heretofore unnoticed lake swells. Within thirty minutes I was positioned next to the floating bog-island, clicking away in hopes of catching a decent shot.

Of course, I wasn't going to stop birding just because I had reached another milestone—700. I was headed to the Pribilofs that week and with any luck I would see the recently reported Code-3 Tundra Bean-Goose.

Kenai NWR, Alaska

Aleutian Tern: *number 700 (Headquarters Lake, Alaska)*

Note: the white forehead is the distinguishing field mark.

31. Cape May Code-5

I was dutifully at my computer reading and responding to e-mails, when my pulse quickened upon seeing the latest NARBA post. Cape May, New Jersey, had just reported a Whiskered Tern. Common in Africa, this bird has been seen only twice before in North America, both times at Cape May (1993 and 1998, and I was living in Ohio and England on those dates). It is so rare that neither my iPhone's bird app nor the Sibley bird guide contained an entry for it. Admittedly, I had been pretty content reveling in the satisfaction of May's trip to the Pribilofs, where I reached 702 North American lifetime birds. I had passed up several opportunities that summer to hop an airplane and try to see another new one. But this Code-5 was almost in my backyard. I took a calculated risk that this mega-rarity would hang around for a few days like the two in the 1990s had.

Whiskered Tern *(Cape May, New Jersey)*

Within seconds of reaching the Cape May Hawkwatch platform, I spotted the bird, as well as several birders from my past, including the Cornell student I'd met two years earlier in the Pacific Ocean (Half Moon Bay pelagic). He had taken a 3:00 a.m. bus from Ithaca to Bergen County and convinced his non-birding parents to drive the four hours to Cape May. I was pleased to show them the distant tern, acting like a swallow, catching insects off the surface of the pond. Number 703 was in the books.

32. Rio Grande Valley

A snowstorm ushered in Philly's first day of spring, but it did not prevent USAir from whisking me to McAllen, Texas, in search of two Code-4s. You will recall that a Code-4 is a once-every-ten-years occurrence. For two days I looked, listened, cursed, and prayed (that was mandatory down there) but unfortunately missed the Gray-crowned Yellowthroat and the Blue Bunting. It was some consolation that eBird, Cornell's birding database, failed to receive any reports of these birds that weekend. Maybe they returned to their home country, a mere twenty miles south.

eBird did report the presence of two Groove-billed Anis on Saturday at Laguna Atascosa National Wildlife Refuge. This bird had eluded me because it is primarily a summer resident in the Rio Grande Valley and its population had been in decline.

I drove to the precise location indicated on the report and waited for sunrise. Two bobcats and the Cameron County sheriff eyed me as I walked in circles around the refuge entrance. After only an hour or so, I noticed a flock of Red-winged Blackbirds flee from a flooded field. Oddly, one remained behind. I took a casual look with my binoculars and was pleasantly surprised to focus on the ani. I walked toward it, and it flushed to a nearby bush. I approached with my iPhone camera and snapped one shot as it flew.

Groove-billed Ani *(Laguna Atascosa National Wildlife Refuge, Texas)*

33. Bird-A-Thon

Spring migration in the Northeast is always exciting. In an effort to benefit the Hitchcock Center for the Environment in Amherst, Massachusetts, I signed up for a one-day birding marathon with my brothers-in-law, Dano and Joe Weisbord.

We birded from 4:30 a.m. to 9:00 p.m., covering 120 miles by car and over twelve miles on foot. Our goal was to positively identify ninety species, and we tallied eighty-nine. Darn. If we'd counted the distant gulls we saw at one point but couldn't absolutely specify, or added the moose (a true rarity). . . oh well. Given that we were limiting ourselves to Hampshire and Franklin Counties in central Massachusetts (so, basically no waterbirds like sandpipers or ducks to pad the list), our total was pretty respectable.

By 7:00 a.m. we had listed thirty species, including the tricky Bobolink, despite an intermittent drizzle. The Silvio Conte River Trail produced many colorful birds, like Orchard Oriole, Blue-winged Warbler, and Rose-breasted Grosbeaks by the handful. It was worth walking to the summit of Mount Holyoke (almost 1,000 feet) as we saw the rare Cerulean Warbler and uncommon Worm-eating Warbler. At 11:00 a.m. our total was fifty-seven, and we hadn't yet seen the ubiquitous House Finch or pigeon.

Groff Park, Norwottuck Rail Trail, and Quabbin Reservoir were a bit too quiet, so we hit recommended spots on the Connecticut River, hoping to boost our total during the midday birding doldrums. We combed the "Honey Pot" and located single Vesper and Savannah Sparrows. A solo Bufflehead (I'll admit publicly that I swore it was a Hooded Merganser and was proved wrong by Dano) and a Bald Eagle resided in Turners Falls as we hit eighty species at 5:45 p.m. Rock Dove (aka pigeon) made eighty-one. Ninety species was still possible but iffy given that the sun was going to set at 8:00 p.m.

The nearby Montague Plains (elevated glacial sandflats) quickly offered up a Ruffed Grouse, Pine Warbler, Field Sparrow, and Tennessee Warbler. A careful summation of the bird list showed eighty-seven species. I had forgotten to add the Northern Mockingbird and Eastern Bluebirds seen earlier. Ninety was now better than iffy. We double backed to the big body of water in Barton's Cove with hopes of squeezing out a gull and a cormorant in the fading light, but none were in sight.

We returned to the eerie Montague Plains, where we were showered with Eastern Whip-poor-will calls from every direction at dusk. A bonus was several American Woodcocks "peenting" at our feet. Eighty-nine. We did not hear any hoped-for owls as we headed home, exhausted and satisfied that we had tried our best and completely enjoyed this birding marathon.

34. In Honor of Shel Severinghaus

My friend Shel Severinghaus passed away after a difficult bout with cancer. As I mentioned earlier, the Severinghaus family was instrumental in developing my interest in birds, starting in the late 1960s with trips to southern Florida. Well lucky me, my wife's employer, Temple University, qualified for a football bowl game in Boca Raton, so I "had" to join her down there. Even better, that elusive Western Spindalis had been seen daily in a Fort Lauderdale–area park since Thanksgiving.

The rare bird alerts made it seem like this bird would be easy to locate. However, the alerts didn't mention that the park was a dog park, a seriously popular dog park. The combination of playful pups, overhead jets, and I-595 traffic made it nearly impossible to listen for or locate the family of Spot-breasted Orioles with whom the smaller, shy spindalis tended to associate. After two hours searching a ten-acre area by myself with no other birders in sight, I was beginning to feel the pressure of my promise to be back by lunch. Fortunately, a birder from Minnesota and another fellow from Tampa showed up. We quickly divided up the territory and within minutes spotted the orioles.

Spot-breasted Oriole *(Markham Park, Florida)*

I had not seen a Spot-breasted Oriole since the one that nested in Shel's father's Coconut Grove, Florida, yard in 1969. And just as promised, the orioles were a true harbinger for North American life bird number 707. A beautiful male Western Spindalis peeked out of the foliage for a moment, and I shot off a barrage of digital ammo. A spindalis in honor of Severinghaus—truly fitting.

Western Spindalis *(Markham Park, Florida)*

The spring break plan was always centered around seeing our Miami-based Sarah, the person who originally suggested my birding blog. A trip to the Everglades was sandwiched between exquisite South Beach dinners, retail therapy, and even Bikram Yoga. Chasing rare birds was honestly not on the agenda, but Sarah's business trip and Nina's encouragement presented me with an opportunity to race down to the Florida Keys at 4:00 a.m. Two specialties had been consistently seen since mid-February in the same state park that had produced lifer number 704 (Key West Quail-Dove) the previous March.

Both species are primarily West Indies natives; the Zenaida Dove is a Code-5, and the other mini sparrow-like Black-faced Grassquit is a Code-4. Seeing both on the same day is like a baseball player turning an unassisted triple play *and* hitting a game-winning home run in the same game.

With directional assistance from a man who had driven twelve hours from Charlotte to guide his two Sacramento friends, I hastily walked a few hundred yards along the Golden Orb path and stopped at the previously marked pink streamer spot in the mangroves. We waited less than a minute and gasped expletives as the dove emerged from the jungle onto the sandy path and fed at our feet. The white patch distinguishes this bird from any other dove.

Zenaida Dove *(Long Key State Park, Florida)*

The grassquit was going to be more of a challenge. Besides being only four inches, drab in color, and quiet (it was a female), it had not been seen in three days, and that was in a private RV area. Add the uncooperative park ranger who did not appreciate our presence in his campground full of paying customers, and we had a recipe for disappointment. Three hours elapsed, and our group grew to more than twenty as folks from Michigan, Texas, and New York City joined the hunt, but there was no hint of the target bird.

I wasn't going to quit on the grassquit, but I did separate from the group to grab my sunblock from the car. As I applied the lotion, I could interpret the body language of excited birders 400 yards away. A short jog at a moderate pace brought me to the spot thirty seconds after the bird was flushed into a tangle of who knows what. Turning away in disappointment, I saw a little olive-gray thing sitting on a water hose contraption next to an RV. Before I could raise my binoculars, the Californians raised their voices, and I fired away with the digital zoom and prayed this was our bird. It was a home run. The score now stood at 709 North American lifetime species.

Black-faced Grassquit *(Long Key State Park, Florida)*

35. Texas and Arizona

The first sighting of an Amazon Kingfisher in North America occurred in 2010 in Laredo, Texas. A second one appeared in 2013 near Harlingen, and then a third (female) frequented the Rio Grande Valley near Laredo in 2017. This bird has earned its Code-5 status. It had been ten months since I'd added a new bird to my North American life list, so chasing this special beauty over the Martin Luther King holiday weekend was the perfect antidote for my nature-deficit syndrome. Only nine hours spanned waking up in Philadelphia and joining a fellow who drove all night from Wyoming on the banks of the Rio Grande.

The bird was stationed exactly where the rare bird reports had indicated. The problem was it was comfortably positioned on the Mexican side of the river, and birds spotted in Mexico cannot count toward one's North American list, according to the American Birding Association.

Silly, I know, but the primary boast among most birders is one's North American life list. Of course, it was a new bird for my world list, and that number should be the true benchmark for who really rates in the birding-world pecking order.

Anyway, birders are an honest bunch, and we follow the rules of the game. We waited. A couple of hours. Just talking and sharing highlights of our birding obsession, all the while watching the kingfisher nab little fish from its Nuevo Laredo perch. Why would this bird emigrate to the United States when the fishing was so productive in Mexico?

Thankfully the answer was a pesky Green Kingfisher. It startled its much bigger cousin, and both flew over to our side and perched on a piling. Now I could count the Amazon—number 710—and contemplate my next move.

I wrote about the kingfisher from my Super 8 motel room in Willcox, Arizona. A long but satisfying day. Philly to Dallas to Laredo to Dallas to Tucson to Willcox. Thank you, frequent-flier miles. Southern Arizona was hosting three species that I'd never seen.

Amazon Kingfisher *(Rio Grande, Laredo, Texas)*

Streak-backed Oriole *(San Simon, Arizona)*

Birding Arizona is always a treat, but the winter season tends to bring certain rarities to the region. For several weeks I had read about a Code-4 Streak-backed Oriole regularly appearing at a remote homestead's orange-slice feeders on the Arizona-New Mexico border. Despite its supposed reliability, there is always a chance of missing any bird. But this immature male was very cooperative, even with a javelina underneath it.

Javelina *(San Simon, Arizona)*

Pyrrhuloxia *(San Simon, Arizona)*

The owner of this property takes care of the birds and travel-weary birders, as evidenced by the numerous feeders and neatly arranged lawn chairs. Other gorgeous species that deserved digital zoom attention included the Pyrrhuloxia, Gambel's Quail, Black-throated Sparrow, and Woodhouse's Scrub-Jay (a recent split from the Western Scrub-Jay that, while not rare, was a new bird species for the life list.)

Gambel's Quail *(San Simon, Arizona)*

Black-throated Sparrow *(San Simon, Arizona)*

Woodhouse's Scrub-Jay *(San Simon, Arizona)*

The Code-3 Rufous-backed Robin is another Mexican thrush that typically appears in Southern Arizona in January. Two of these had recently been reported a couple of hours away from where I was but in somewhat different directions. I chose the western location (Rio Rico). Upon arrival, I met a Dallas birder who claimed this secretive, non-vocal visitor was not being consistently seen in any particular area of "the spot." Bad choice.

With three hours of light and two hours to drive to the other spot, north of Tucson, I decided to make a move and pursue the Rufous-backed Robin, which apparently likes pomegranates. Set back on private property, the deciduous fruit-bearing shrub was easy to see with a telescope. And the birding gods once again rewarded me. A blurry 500-meter telephoto shot but no doubt this is *not* an American Robin. Can you tell the difference?

Rufous-backed Robin *(Pinal, Arizona)*

Postscript: As excited as I was in my own little birding adventure, the big news for any birder that year was the discovery of a Ross's Gull in Half Moon Bay, California, that Friday. Probably another victim of climate upheaval, this true Arctic bird has basically been seen in the ABA Area only in Barrow, Alaska, in the fall. Confirmation of multiple Saturday sightings via social media gave me no choice but to alter my air tickets. Southwest Airlines is very cooperative. I prepared to leave for San Francisco in twelve hours. At 10:00 p.m. the night before, I checked the San Mateo rare bird reports again to make sure—I don't know why, OCD I guess. And I discovered both in words and photos that the previous afternoon the Ross's Gull had been "predated" by a Peregrine Falcon! I headed home.

36. Tupper Lake

It makes no sense, but another Ross's Gull was reported in the Adirondack region two weeks later. I don't believe two Ross's Gulls had ever been spotted in the United States (excluding Alaska) in the same year. This incredible sighting was clearly a result of climate change combined with a greater number of keen-eyed birders around the country. This first-year-very-lost male had been walking on frozen Tupper Lake (near Lake George) for several days desperately searching for its next meal. Locals tossed carrion onto the ice to help it refuel. It was likely that another Peregrine would predate this little, fatigued, and conspicuous gull any day.

Fortunately for me, I had a trip planned to Vermont that week. A three-hour detour to see a life bird was a no-brainer. But would the bird survive long enough for me to see and photograph it? Selfish thought, I know, but that's being totally honest.

As I pulled into the designated viewing spot, the Tupper Lake boat launch, an enthusiastic young fellow dressed in barely more than a tee-shirt and blue jeans in the 10°F temperature waved me over to a group of ten others gathered at the edge of the ice. I checked the license plates of the other cars as I tried to calmly walk over to them. Connecticut, Massachusetts, Maryland, New York, New York, New York, and me. The bird was about a quarter of a mile out on the ice, and the twenty-six-year-old Yale Ph.D. student eagerly showed me the Ross's Gull through his scope. I thanked him, learned he was at Middlebury at the same time as my son Tyler, and he was quickly on his way back to New Haven to make a night class.

I reflected on our brief exchange. Seeing another life bird is exhilarating, and it is comforting spending moments like these with my community of birding zealots. But mostly I was gratified to know that the next generation of birders had fledged.

Ross's Gull *(Tupper Lake, New York)*

37. Wild Oriole?

Berks & BEYOND
COUNTY, STATE AND REGION

Rare bird sighting causes frenzy

Okay, maybe not the *The New York Times*, but the *Reading Eagle* pictured the unprecedented Black-backed Oriole on the front page of its February 4 edition. Why? Because this Central Mexican bird had never been reported in the United States or Canada, and one was frequenting a birdfeeder an hour from Philly.

There was no doubt that this bird was a Black-backed Oriole. The real question was whether it was wild or an escaped cagebird. When I'd first heard about it, I completely dismissed the notion that it was wild. But I kept reading daily eBird reports from more and more birding luminaries who were trekking to Berks County to spot what could be an ABA first. It was in perfect condition, atypical of pet birds, which tend to have damaged feathers. My friend Brant Henderson reported seeing a television news story: why hadn't I seen the bird? For two weeks this colorful but shy male visited fresh orange slices put out by a friendly couple who were enjoying their ornithological fame.

Black-backed Oriole *(Reading, Pennsylvania)*

I had a flashback to the summer of 1983 when I failed to try to see the now-infamous African Western Reef-Heron on Nantucket, while I was spending a weekend on nearby Martha's Vineyard. That first-ever North American sighting still haunts me, as I didn't believe I could afford the extra vacation day and cost to chase it. I vowed to never make that mistake again. An hour's drive to see a beautiful orange, black, and white bird in the suburbs of Reading on a dreary February morning was certainly worth my time.

The guest book had over 1,000 names of visitors from twenty-six states and three countries. This bird was already a national celebrity in the birding world. It took a year for the authorities to rule whether it could count as wild or not. It counts.

38. Persistent and/or Stupid

Visiting our friends Marion and David Knowles in San Diego was the primary goal of our western trip. Unfortunately, I was struck hard with a nasty cough at the end of the weekend stay. Rather than catch a plane home with Nina Monday morning, I stubbornly stuck to my plan to chase three lifers. The Gray Vireo reported that week by eBird, only seventy-five miles east of San Diego, proved to be an easy spot. The mile-long hike at sunrise at an altitude of several thousand feet did not help my developing pneumonia, and my camera fell ill, as well. No photograph.

The smart move would have been to have seen a doctor in Philadelphia by now. But no, I had a round-trip boat ticket to Santa Cruz Island in the Channel Islands National Park, and I was not going to miss my opportunity to see the Island Scrub-Jay in the only place in the world it resides. It is related to all jays, but it is its own unique species that never leaves the island. Another strenuous hike led me to five of the 12,500 members of this species that call Santa Cruz Island home.

Island Scrub-Jay *(Santa Cruz Island, California)*

Of course, I should have been en route to *any* doctor by now. But no, the Code-4 Tufted Flycatcher was a major rarity that was going to be relatively simple to find once I boarded a flight to Tucson and drove 105 miles. My persistence was rewarded, but admittedly, I really did not enjoy the experience. I just wanted to catch up on sleep that pneumonia's persistent cough prevented.

Tufted Flycatcher *(Cochise, Arizona)*

Birding-wise I was very proud of closing in on another milestone—723 ABA Area lifetime species—but I was also a bit embarrassed by the health risk I took in the name of chasing birds.

There would be more stories, but I promised to listen to my body, as all of us should, and make the correct decision next time.

39. Three-Island Tour

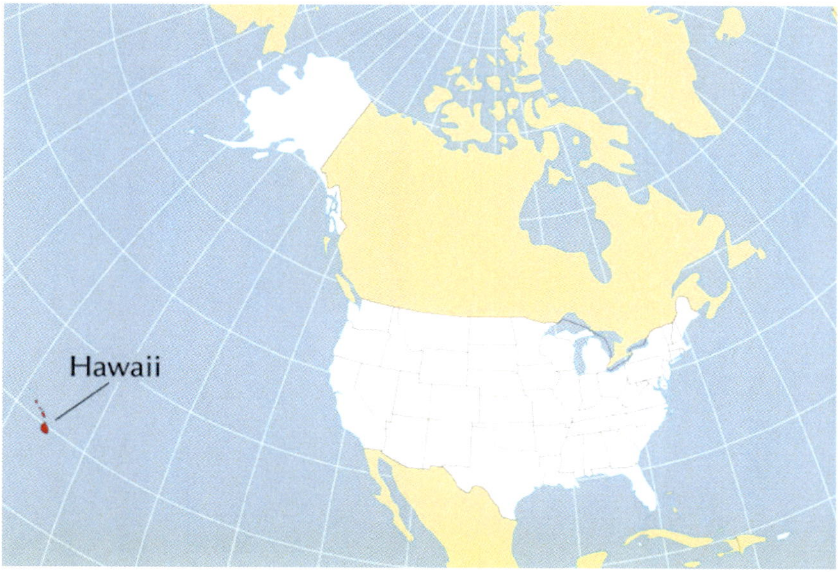

Earlier, I specifically mentioned that Hawaii was not a part of the official ABA Area until 2016. Who knows why that was the case; it has been our fiftieth state since 1959. The Hawaii Department of Tourism finally won its battle with the birding powers, and now Hawaii birds can be counted on one's ABA life list. More than 350 bird species have been recorded in the eight major Hawaiian Islands, including thirty-four endemic birds found nowhere else on earth. The goal of 800 ABA Area lifetime species was now in play.

A knee replacement in August 2017 had kept me on the bird-chasing sidelines for six months. But a favorable March work schedule, coupled with the blessing of the home-front medical team, set the table for what was a remarkable three-island tour in Hawaii. Yes, I admit that I went to Hawaii while my very understanding wife remained home in Pennsylvania. The fact that Nina has two birding brothers helped my case.

Hawaii is an archipelago with super-rare birds that can be seen only there. And sadly, due to climate change, several species will likely become extinct in my lifetime. Why? Because malaria-carrying mosquitoes are moving up the hillsides as temperatures rise, and the birds—most of whom never developed an immunity to these diseases—are literally flying out of room seeking lower temperatures at higher altitudes.

The lucky ones that seem to have some resistance to avian pox are threatened by introduced ground predation in the form of feral cats, pigs, and mongooses.

My gifted guide, Mandy Talpas, created a fascinating itinerary that included time on three islands: Oahu (home of Honolulu), Hawaii ("the Big Island"), and Kauai ("the Garden Isle"). We found sixteen of the seventeen possible endemic bird species (we missed the Hawaiian Hawk) and ten of those sixteen are listed as endangered by the U.S. Fish and Wildlife Service.

Hawaii has native specialties like the Hawaiian Goose (Hawaii's state bird).

Nina Weisbord

Hawaiian Goose, "Nene" *(Kauai, Hawaii)*

Hawaii also hosts a number of introduced species like the Red-crested Cardinal (from South America). Most of these birds (from Asia, Africa, South America, you name it) have been accepted by the birding authorities at the ABA, as well as by the American Ornithological Society (AOS).

Red-crested Cardinal *(Oahu, Hawaii)*

Pueo *(Kauai, Hawaii)*

The Pueo, or Hawaiian Owl, was an excellent find. It is a subspecies of the Short-eared Owl, and as a subspecies, it doesn't count as a new species for listing purposes.

I had a realistic potential of seeing fifty lifers on this trip. I added forty-six more. And my total stood at 771.

40. Bob's Woods

My heart jumped when I read the New Jersey Rare Bird Alert text on Sunday morning April 29, 2018. The Swainson's Warbler had eluded me in Florida and two Carolinas, and now it was only two hours from Philadelphia, hundreds of miles north of where it was supposed to be. By 2:00 p.m. I was anxiously wandering the footpaths near the Cape May Point lighthouse hoping I had interpreted the eBird coordinates correctly, when I ran into my birding friend Win Shafer. Together we determined exactly where this small (five-inch) bird had been heard but basically not seen. Unfortunately, after a couple of hours standing in the bird-sound-less wind, we each headed home, very disappointed.

For the next two weeks I jealously read daily reports of birders hearing but rarely seeing this same singing male. This bird must have set up a breeding territory; otherwise it surely would have moved and continued its wayward migration. The Swainson's Warbler is an olive-brown bird with a huge voice. Think Carolina Wren. When not singing from an exposed branch in breeding season, it spends most of its time skulking on the ground, hidden in dense thickets, and foraging on the ground looking for millipedes, spiders and caterpillars.

This guy was particularly difficult to locate because it was situated deep in the labyrinth of private property.

On Monday, May 14, I woke up very early and decided to try again. Fueled with Wawa coffee, I cruised to southern Jersey and was planted in "the spot in Bob's Woods" just before dawn. The 3:00 a.m. wake-up was worth it, because the bird was already belting out its distinctive song very loudly. It wasn't that close and, of course, it was hidden somewhere in the distant dimly lit stand of trees. I desperately wanted to see it, since I don't count heard-only birds on my life list. I was alone.

Should I play the bird's song on my iPhone to draw it closer? I thought of the occasional golfer with a bad lie in the rough who moves the ball to ostensibly check its identity and replaces the ball in a better position. That's called cheating. No, using the phone during breeding season would be unethical and kind of cheating. It would be very disturbing to this persistent, lonely male who has already been exerting tremendous vocal energy for fifteen days to impress a non-existent female.

At 6:15 a.m. I noticed a slight silhouetted movement twenty feet up and about fifty yards away. I was rewarded with a 10x binocular-aided visual of this Code-2 wood-warbler. It was singing from a branch in a tree that had not completely leafed out. Eureka. I reached for my camera. Oh, right. I didn't think I had a prayer to photograph this bird so I'd left the camera and its telephoto lens at home. But I did take an iPhone shot of the tree it occupied.

You'll just have to trust me. ABA Area species number 772 and the last of the fifty warbler species (excluding Mexican rarities) one can see in the Lower 48 states.

Swainson's Warbler *illustration from National Geographic Field Guide to the Birds of North America*

41. Arizona (again)

Adrenaline pumping, I quickly traversed down the steep and narrow trail back to my rental car. After patrolling the stakeout area for three hours, I was certain I had just captured an image of the distant target bird. As I examined the photo in the shade of the front seat, my heart sank. There was white in the wings. Ugh. I had seen the common Painted Redstart rather than the Code-4 rarity Slate-throated Redstart.

I had traveled to southeastern Arizona for an encounter with this Mexican specialty. It had been seen daily but sporadically for a couple of weeks. I re-read the recent eBird postings describing the habits of this small warbler and headed back up the ravine. Sunset was my only deadline.

I sat at the end of the 200-yard trail by a little waterfall and waited. Within minutes, two birds caught my eye as they darted among the conifers and oak trees, mimicking each other's moves. I had a good look at the lead bird. It was another male Painted Redstart. Prior birders had mentioned that the male Painted and the female Slate-throated appeared to be exhibiting courtship behavior.

This pair continued their coordinated but randomly patterned flight dance. I struggled to focus my binoculars on the second bird. They took a break and perched in a viewable spot twenty yards away. I had my Slate-throated Redstart.

You didn't think I'd trek to southeastern Arizona in pursuit of only one rarity, did you? Nah. There were *two* birds on my hit list. Only eighty miles from the Slate-throated Redstart, in Ramsey Canyon, a Code-3 Flame-colored Tanager had been seen on and off for about three weeks. This is an area over 6,500 feet in elevation with a uniquely cool environment unusual in the desert Southwest. It's the same nature preserve where I unknowingly had pneumonia and unwisely tracked and photographed the Tufted Flycatcher in 2017.

Slate-throated Redstart
(Pinery Canyon, Arizona)

Flame-colored Tanager
(Ramsey Canyon, Arizona)

Remember my 2013 encounter with a hybrid Flame-colored x Western Tanager? A *yellow*-orange bird but not countable because it is not considered a separate species. Seems odd to me that I couldn't count it as anything and there it was. Oh well, the birding authorities rule.

Anyway, the Flame-colored Tanager (a *red*-orange color) was being seen along a half-mile stretch of the Hamburg Trail. This rocky path wound through dense stands of conifers, oaks, and sycamores. Viewing would be difficult, as this medium-sized passerine prefers to glean insects from the foliage in the mid-to upper levels of the trees. I arrived at Ramsey Canyon Preserve and happily paid the $6 fee. Daylight was not my limiting factor; rather the 5:00 p.m. closing time imposed by The Nature Conservancy. I did not want to be on the wrong side of the iron gate when the staff locked up.

After three hours of great views of colorful Red-faced Warblers and Hepatic Tanagers, among other not so exciting interruptions by common woodpeckers and jays, I found a spot where I had an unobstructed ninety-degree view of the tops of the trees in this expansive territory. It was 4:00 p.m. I estimated I was thirty minutes from the gift shop, or, put another way, I had thirty minutes to find this bird. The sun was casting long shadows in the canyon. I stood there beginning to rationalize the Slate-throated Redstart sighting earlier that day as sufficient justification for spending three days (and a bit of money) in Arizona, when I heard a distant call that sounded like our eastern Scarlet Tanager.

I binoculared the farthest sunlit treetops, and a bright red figure outlined in coniferous green stood out. I fumbled for my camera, steadied my hand, and clicked away. The magic of zoom lenses and automated digital photography never ceases to amaze me. The Flame-colored Tanager disappeared after only a minute or two.

As I hustled back to the parking lot I was greeted by a curious Greater Roadrunner. The state bird of New Mexico, the roadrunner is a long-legged member of the cuckoo family commonly found throughout the Southwest.

Greater Roadrunner *(Ramsey Canyon, Arizona)*

42. Texas (again)

"Hi, Laura . . . Ohio." "David, Minnesota." "Leslie, San Antonio." "Barbara, Lauderdale." "George, Philly." These were the brief but polite exchanges as we greeted the young Ph.D. student from Corpus Christi. None of us wanted to engage in actual conversation for fear of missing a glimpse of the Mexican Violetear, a hummingbird that rarely emigrates from central Mexico to the United States. The six of us stayed positioned in one spot at Quinta Mazatlan, a Spanish Revival–style mansion with gardens—an urban sanctuary. We were desperately hoping for the appearance of a metallic-green bird formerly known as the Green Violetear that had made daily trips to the purple and red flowers in front of us for the past thirteen consecutive days.

Having nailed my two Arizona target birds (both on Father's Day) with a vacation day to spare, hitting South Texas on the way home was logical. For David, the twenty-four-hour drive from Minneapolis earned him the right to occupy the pole position, the location along the short fence with the widest view of the clearing. Laura and Leslie had had five-second looks at the bird the day before, so they were content to stand behind Barbara, David, and me, cameras pointed at the mimosa bush. I never did catch the Ph.D.'s name and frankly didn't care. We waited four hours until closing time. Buff-bellied and Black-chinned Hummingbirds but no *Colibri thalassinus*.

The next morning, the "mansion with a mission" opened at 7:00 a.m.—an hour early for us birders. The same group congregated and waited four more hours. My last possible flight out of McAllen required that I leave at 11:00 a.m. Disappointed on one hand but also satisfied with my effort, I made it to the airport with minutes to spare, having failed to allow for the extra wait time required to clear Border Patrol's checkpoint. I was in McAllen, Texas, the focus of national attention on human, not avian, immigration. The topic never came up with my fellow self-absorbed birders, but the irony of the situation didn't escape me.

No words needed

43. Perseverance

***Black Rail** illustration from National Geographic Field Guide to the Birds of North America*

You will recall that the Black Rail is one of, if not the most difficult bird in the ABA Area to see. Or even glimpse. It is super small and secretive. I am confident that I heard but did not see the elusive Black Rail in the Corte Madera Marsh in 2014. It would have been number 700, but as you know, I do not count heard-only birds.

eBird was reporting several of these shrew-sized birds reliably calling during the day (normally an evening activity) in the Alviso Marsh in the Don Edwards National Wildlife Refuge near San Jose. Since I was already in San Francisco, the fifty-mile drive south was a non-decision.

Alviso Marsh

I arrived at dawn, and two other birders were already in place at the end of the boardwalk. We heard several Black Rails give their signature *kic-kic-kerr* calls. Seeing one of these shy, tiny marsh-slinkers is considered one of the most difficult achievements in North American birding; photographing one is nearly impossible. After two hours the guy gave up. The woman birder lasted another two hours but also departed, leaving me by myself.

The calls were coming from only a few feet away, but I could not see any movement in the dense marsh grass. I had to leave in forty-five minutes to make my business meeting in Mountain View. My frustration was peaking when I saw something walk by an empty Gatorade bottle. My heart jumped as the bird gave me a two-second look. It was a rail, but it was the much more common Virginia Rail. But wait, another shadow caught my eye as I peered down over the railing. It looked like a mouse, but it wasn't a rodent. The Black Rail! The telltale white flecks dotted its back, and a maroon nape gleamed in the bright sun. I watched it for at least five seconds, and then it was gone.

Incredibly satisfied—I mean so content I cannot explain—I drove to my meeting in my birding clothes. Thankfully, the fellow I met was quite understanding and, it being California, he didn't expect to see me in a coat and tie. My next meeting had cancelled earlier, so I had a crazy thought. Why not go back to the marsh and attempt a photo with my iPhone? Of course, I had not even considered bringing my Nikon camera on this trip. Nobody photographs a Black Rail.

I returned to the boardwalk and waited with my iPhone camera on, arms extended over the exact spot I had earlier seen the rail. It was 2:00 p.m. Every ten minutes or so a rail called. Clearly the end of the boardwalk was in the middle of its territory. I waited and waited. At one point I returned to my car and recharged my phone for fifteen minutes. By 4:00 p.m. the calls were more frequent. I prayed, and I questioned my sanity. But my patience and persistence paid off.

Black Rail *(Alviso Marsh, San Jose, California)*

I caught a red-eye flight home that night enabling me to meet Chloe and Jimmy Bereolos's three hour old arrival—grandbaby Areli—at Bryn Mawr Hospital the next morning.

44. South American Hawk

The Great Black Hawk had *never* been spotted in the United States before 2018. Its normal range is South America to Mexico. The fact that one was being seen periodically in Maine that fall was simply astounding. Since Thanksgiving I had been keeping a close eye on the rare bird alerts. Two close buddies of mine (Haden Gerrish and Jay Pitocchelli) had seen the bird, so with the help of American Airlines, I made my move on a Saturday in mid-December.

By 10:30 a.m. I was in Deering Oaks Park. Several birders had been there since dawn; no sign of our Code-5 visitor. The winds were calm, but the temperature was 19°F. The park is host to a healthy gray squirrel population, a very popular food source for most raptors. The Great Black Hawk had reportedly feasted on at least one of these unsuspecting creatures daily.

After a couple of hours of pacing around the park, I noticed a raptor in a nearby tree. For a moment I thought the birding gods had smiled on me once again. But no, it was a Red-tailed Hawk. More than five hours had elapsed, and I couldn't feel my fingers or my toes. Adrenaline had sustained me until then. I was frozen. The sun sets at 4:03 p.m. up there. I was done. This is not a triumphant story about ABA Area Lifer number 776. Rather, it's a chronicle of one day in my life when my patience was tested, and the subsequent frustration had to be conquered by the fact that trying is more important than the end result.

But of course, I had to try again to see and photograph the first-ever-documented-in-the-United-States Great Black Hawk. So, two weeks later, December 26, with the aid of Nina's frequent-flier miles, I was back at Deering Oaks Park, accompanied by at least thirty avid birders from around the country. The picture tells it all.

Great Black Hawk *(Portland, Maine)*

Sadly, this bird eventually suffered from frostbite and was picked up by a wildlife rehabilitation clinic. A few days later it was euthanized.

45. Siberian Flycatcher

Red-flanked Bluetail *(Los Angeles, California)*

The birding world was rocked by the news of a Code-4 female Red-flanked Bluetail, an Old-World flycatcher, inhabiting an enclosed library courtyard at UCLA. I was already booked on a Sunday flight to Los Angeles for work. Truth be told, I had never even heard of this species until an Idaho birder reported one two years earlier. They winter in Asia—in the Himalayas and Indochina. As I approached 800 lifetime ABA Area birds, this one was not on my hit list.

The *Los Angeles Times* ran an article on Saturday about the masses of birders descending on this oasis in the city and the difficulty people had seeing, never mind photographing, this skittish thrush-like creature. A winter downpour commenced at five o'clock Monday morning and continued past dawn. The crappy weather did not temper my excitement but did force me to buy a cheap umbrella and a lousy poncho at a CVS.

I showed up an hour earlier than the posted 9:00 a.m. opening time and snuck into the grounds behind an employee with the magic gate-opening swipe card. But as the big metal gate shut me into the several-acre William Andrews Clark Library grounds, I had a slight panic that maybe I would not find a five-inch bird in a much bigger area than I had anticipated, especially in such a downpour.

Consistent sightings had been reported near a ficus tree and a U-shaped hedge. A what, where? Damn, I couldn't get cell service to even google a ficus tree. Thankfully, after an hour, I was joined by retired police officer Jim, who had flown home to San Francisco from Argentina the day before and then driven all night to chase this bird. He knew exactly where we should look. Within minutes, an LBJ flashed in front of us and posed on a (ficus?) branch. The orange-red flank and iridescent blue tail were evident, even in the rain.

46. Southeastern Florida

The description of a Thick-billed Vireo reads a little like an LBJ. "Medium sized (5–6 inches), olive-green upperparts, brown iris, pale yellow underparts. Most often heard, not seen. Frequents dense undergrowth and thickets."[5]. Oh, and it's a Code-4—meaning it pops over from the Bahamas to southeastern Florida, maybe once every five to ten years. Frankly, I never felt I'd be in the right place at the right time to find this non-migratory bird. Therefore, I never expected to add this accidental United States visitor to my life list. Until Presidents' Day.

Fortunately for me, I was traveling to Fort Lauderdale on a 5:45 flight Monday morning, a day after the rare bird report texts came over my phone. Crandon Park, Key Biscayne; totally doable from Lauderdale. Everything went smoothly, and I pulled into an empty Crandon Park Nature Center parking lot at 10:00 a.m.

My early concern that this was going to be another "needle-in-the-haystack" search was alleviated as I was quickly joined by a woman from Orlando, a man from Milwaukee, a father-son duo from Toronto, and a big year participant from Australia. Within minutes our target vocalized, and we caught glimpses of him darting through the dense undergrowth, hunting for insects.

I had my lifer, but a photograph seemed impossible. The others also were equipped with cameras, but the conditions were not permitting a decent photograph. Over the next three hours, perhaps satisfied with—at best—a dark silhouette image of this passerine, the others wished me well and departed. Except for the very persistent and friendly Canadians. Not much longer after the self-described "impatient ADD" man from Sydney succumbed to the bugs, the Toronto guys motioned me through the thickets to come to the courtyard. There was our bird, gleaning a meal from an eye-level bush. I think you will agree that he is far from drab or dull and rather spectacular.

5 iBird Plus app

Thick-billed Vireo *(Key Biscayne, Florida)*

47. Garganey

Miles Horton and George (Sacramento, California)

We were at the Palace Hotel in San Francisco for a fun-filled Parish family wedding extravaganza weekend, and I had no intention of birding. We had three packed days of celebrating the newlyweds and catching up with friends. I did not bring my Nikon camera, but I always travel with binoculars. I figured I might find an hour to view the amusing parrots of Telegraph Hill near Coit Tower.

On Saturday afternoon I received a Code-4 text alert. A male Garganey was dabbling with Cinnamon and Blue-winged Teals in a flooded field near Sacramento, far away from its Eurasian-Aleutian range. I had an adrenaline rush but quickly dismissed any chance of chasing this fellow on Sunday because I thought Sacramento was way out of my range. And I simply had to return to Philly sometime Monday.

Our Sunday night dinner hosts, the Hortons, politely listened to one of my birding monologues at cocktail hour and questioned why I wasn't going to try for it on Monday morning. Their eldest son, Miles, a lifelong friend of our Chloe and Izzy, was joining us for dinner before heading back to his apartment in California's capital, only a ninety-minute drive away. Wait . . . *what*? My geographic knowledge of our third largest state had failed me. I had calculated it was four hours away.

Thankfully, Miles's curiosity about bird-chasers and interest in searching for this aquatic specialty near his office convinced me to adjust my Monday itinerary. After a wonderful meal and catch-up with the Hortons, Miles and I went our separate ways, having agreed to meet at eight o'clock Monday morning at the flooded field.

Within minutes of our arrival, three guys with scopes located our distant target and gleefully shared their views with us. I didn't even need my binoculars. Miles had started his life list with a Code-4 Garganey. To be honest, sparking a possible interest in this hobby in a young person, an environmentally conscious young person working with a state representative, gave me more satisfaction than checking off my ABA Area life bird number 781. That day was a win-win.

48. Concan YG

Grosbeaks are colorful seed-eating passerines with oversize beaks. The United States has several species, but I never thought I'd have the opportunity to see and photograph a Yellow Grosbeak. Only a handful of "YG"s have been documented in Arizona and New Mexico since 1995. They have never been reported in Texas. But a friendly homeowner in Concan, Texas, had alerted birding friends in early January that a large, bright yellow finch-like bird was frequenting her feeder. It had been a consistent visitor for almost three months, and I simply could not resist the chase.

A road sign reminded me that this bird could be an escapee rather than wild, but most felt that the Texas birding authorities would rule in listers' favor. Hundreds of birders who trekked to this remote southernmost part of Texas Hill Country would have been pretty upset if this bird didn't count. Obviously, I was confident, or I wouldn't have spent three days traveling back and forth to Concan.

Our local guide had instructed what turned out to be twelve of us, from all over the country, to meet at Neal's Lodges, where we would consolidate our vehicles and caravan to the secret spot. While we hoped this bird would turn up as soon as we arrived, it doesn't work that way. We patiently waited inside the house and watched the numerous feeders outside in complete silence. An hour passed, and short conversations started but quickly stopped as we desperately tried to turn a goldfinch into a Yellow Grosbeak. Another hour passed, and I spoke to the guy I had been rubbing elbows and hips with for two hours. A wildlife biologist from Charlotte, he had an ABA Area life list over 800. And he hadn't even been to Hawaii. I was humbled. These were serious birders.

Yellow Grosbeak *(Concan, Texas)*

Thirty more minutes passed, and I noticed that I was the last one standing, as the others had taken seats in our generous hosts' dining room chairs. I wondered if this was a minor endurance triumph related to six years of riding Manhattan subways. Or perhaps standing for hours on boats whale watching in the Bay of Fundy had given me an edge. Or was it the Bikram Yoga?

No, I countered to myself. This wasn't a competition. I was simply polite, stupid, or stubborn.

Suddenly my peaceful daydream debate was shattered by several simultaneous exclamations: "That's it!" "On the feeder."

High fives, fist pumps, hugs, and sighs of relief and elation followed. Our hosts and guide smiled. A bunch of strangers united by a yellow bird.

49. Armchair Birding

While the theme of this book is chasing rarities, I don't want to leave the impression that only Code-4 and -5 birds matter. *All* birds matter, because they are an ecological litmus paper. I really do enjoy all species, except maybe invasive European Starlings and nasty brown-headed cowbirds—they don't even deserve capital letter punctuation.

April produced a nice variety of bird species at my own backyard feeders. We had recently moved from urban Philadelphia to suburban Haverford, Pennsylvania, partly because I missed this area's natural beauty. You may be more motivated to take your own birding interest to the next level if you knew that you could see cool birds without getting up from a chair.

So that's what is depicted below. A selection of photos of several species taken from my morning-coffee chair.

Indigo Bunting *(Haverford, Pennsylvania)*

The iridescent blue Indigo Bunting is a stunner that I had not seen at a feeder since 1968 with Jeff Dingle at his grandfather's home in Coconut Grove, Florida. Its presence in mid-April was a harbinger of true spring and migration.

Hairy Woodpecker *(Haverford, Pennsylvania)*

The Hairy Woodpecker is a bigger version of the much more common Downy Woodpecker. It was the sixth woodpecker species I saw in my yard that month.

Carolina Wren *(Haverford, Pennsylvania)*

The Carolina Wren's white eye-stripe distinguishes it from the similar House Wren. If you live in the East, you have likely been awakened before dawn by its loud *teakettle-teakettle-teakettle* song.

Hermit Thrush *(Haverford, Pennsylvania)*

The Hermit Thrush is named for its retiring ways. I was surprised to see one next to the house.

Rose-breasted Grosbeaks *(Haverford, Pennsylvania)*

A male and female pair of Rose-breasted Grosbeaks was a pleasant treat.

After enjoying these local beauties, I had plans to join a dedicated group of birders for a Pacific Ocean adventure. Our goal was to see unusual offshore birds and—we hoped—a blue whale. The chances of decent photographs of my target species (three types of petrels and the world's largest mammal) were very slim. But I hoped to have an entertaining story to share.

50. Pacific Petrels

Star Princess

In mid-May I completed a three-day cruise from Los Angeles to Vancouver aboard the *Star Princess.* Two thousand folks reveled in the round-the-clock gambling, drinking, and other entertainment on this 950-foot-long, ten-story-high monster, while I joined fifty bird-crazed people outside on the bow of the ship from 6:00 a.m. to dark, exposed to the cool Pacific Ocean climate. Why? Because this is the best way to catch a glimpse of fast-flying Pacific pelagic seabirds.

More specifically, the target birds were petrels. Petrels are small ocean-going birds that return to land (generally remote islands) only to breed. They appear to walk on water, à la Saint Peter, as they hover and flutter over the ocean's surface. Our leader was one of the world's premier ornithologists, Paul Lehman. His knowledge is astounding, but even more impressive is his endurance. For two and a half days he never left his feet or took his eyes off the horizon.

We stared for hours across the blue-gray sky and sea, and, except for a few humpback whale spouts, the only signs of life were intermittent appearances of common petrels, shearwaters, and an occasional puffin. Admittedly, I took a couple of twenty-minute respites each day to relieve my legs and back, and yes, some boredom. But Paul had his meals brought to him, and without his skill, we never would have had the ten-second look at the rare Hawaiian Petrel on our second day.

The Cook's Petrel was a target species, but we were enveloped by fog for the few hours when we crossed its territory, and thus we dipped on that bird. Murphy's Petrel was the expected lifer for me, but it wasn't until the end of our second day that one thankfully crossed our bow. If I had rested for ten more minutes, I would have missed it.

Murphy's Petrel *(Pacific County, Washington)*
Photo by Paul Lehman

This trip was a rite of passage if you're going to call yourself a serious birder; a marathon-type experience that I am unlikely to repeat. No regrets: two lifers to bring my total to 784 ABA Area species.

51. Caribbean Visitor

The White-cheeked Pintail is a Caribbean lake-dwelling duck that every once in a while makes an appearance in southern Florida. Lely Resort in Naples had been the home of this presumed wild vagrant for a couple of weeks. Our MLK weekend plans placed me close enough to chase this beauty.

But this bird was reported to spend most of its diurnal hours asleep on an isolated shore of the resort's golf course, hiding its diagnostic facial markings by tucking its head into its body. Finding the bird was one challenge; obtaining an identifying photograph (no, I didn't have my telephoto lens) was unlikely. And the alligators . . .

Thankfully, a friendly fellow with a tripod and scope was visible the moment I turned off I-41 and entered Lely Resort. He tipped me off to the bird's location before I could state my name and impress him that I lived in Philly. Spotting the camouflaged specialty on the shore wasn't a problem because four other bird enthusiasts were already viewing the distant pintail. It was on the rocks in front of the snack bar/restroom, a small building perched on a hill between the ninth green and the tenth tee. But positively identifying this seemingly headless mass of brown feathers was impossible. I needed a closer look.

I waited until the others departed and made my way around the meandering shoreline and approached the snack bar area from behind. My plan was to sneak up on the bird from the side slope of the hill and snap a photo once it reared its head. I was fully aware that I was flirting with violating the birders' code of ethics. I looked back to make sure nobody saw me,

White-cheeked Pintail*, lower right (Naples, Florida)*

but darn, three more birders had taken a position on the far shore and surely had me in their scopes along with the yet undisturbed duck.

I returned to the far shore and greeted the three scopers with an overly enthusiastic: "I had to use the restroom. There was a maintenance man between the bird and me. I wasn't as close as you probably thought. Truncated view, you know," before they could chastise me. I don't think they bought it.

Anyway, without our exchanging pleasantries, they finally left. I repeated my meander. Nobody in sight. No guilt. Lifer number 785, provided the authorities deemed this a wild bird.

52. COVID Common Cuckoo

Common Cuckoo *(Johnston, Rhode Island)*

The COVID lockdown enabled me to concentrate on birding my yard and local areas for most of 2020, and I hadn't chased a rarity since January. The appearance of a Eurasian Common Cuckoo in Rhode Island on November 1 more than caught my attention. Normally this species spends the late fall and winter in sub-Saharan Africa. Clearly this fellow was lost. As it was only the third time this species had ever been seen in the Lower 48 (why it's a Code-3 and not a Code-5 perplexes me), the American birding community mobilized.

I was not able to find ten hours for the round-trip drive to Snake Den Farm in Johnston, Rhode Island, until Sunday November 8. In order to comply with Rhode Island law, I took my COVID test within seventy-two hours of visiting and proved negative.

Thankfully, the weather had been unusually pleasant, and this confused juvenile was content eating the vast supply of woolly caterpillars at the farm. I took my chances and left Haverford at 3:00 a.m. I pulled into the parking area adjacent to the farm at 7:30 a.m. and was surprised to see more than twenty cars with many different license plates already there. I walked briskly toward the assembled crowd a couple of hundred yards away. A friendly couple approached and greeted me with a "Your timing is good." I replied, "It better be, I just drove up from Philly." They smiled and said, "That's nice, we drove fourteen hours from Columbia, South Carolina." This Cuckoo was an attention grabber.

Within minutes I had the stout vagrant in my view. I had planned to meet my brother, Wistar, at 7:30 a.m., but I didn't see him. He lives only an hour away, so I was about to text him when he texted me: "Where are you? I'm here." I looked at the two guys to my right and one was him. I had walked right past him in my excitement to see this anomaly. Blame the masks.

2020 was not only a COVID nightmare but it also cost me a species on my life list. I went backwards by one species when the birding powers decided to lump the Northwestern Crow back in with the Common Crow, as scientists have determined that the Northwestern Crow is a subspecies rather than a stand-alone species. The addition of the Common Cuckoo brought my ABA Area life list back to 785.

But who's counting? I am!

Brother Wistar and George (Johnston, Rhode Island)

53. A First for Pennsylvania

The Tundra Bean-Goose breeds in the tundra of northern Eurasia. It winters in agricultural landscapes in eastern China, Central Asia, and Western Europe. The only reason it is a Code-3 and not a Code-5 in North America is because one can get lucky seeing it, in Alaska.

One morning I was walking our puppy, Lucky, on an extra-long route before an impending snowstorm. I glanced at my phone for the time, thinking I was cutting it close for a scheduled Zoom call. A text interrupted me—oh good, my call had shifted to 12:30 p.m. Just then another text came in from NARBA. It was reporting a Tundra Bean-Goose being seen on the Springton Reservoir in Media, twenty minutes from my house. No way. I had not seen this goose in the Pribilofs in 2014, so it would be a lifer. I reread the text. It was true.

I immediately called my birding buddy Win Shafer, half expecting he had reported the bird. Well, Win was out scouting birds for Saturday's Christmas Bird Count, and he happened to be at Episcopal Academy on Route 252 near Springton Reservoir. But he hadn't seen the alert for this never-ever-been-seen-in-Pennsylvania goose.

Within fifteen minutes I was packing my car with scope, binoculars, camera, and Lucky, when my cell rang. Win was on the bird. "Go to Springton Middle School." I was off. The eleven-mile drive felt like forever. I imagined every car on the road had plans to see the goose. Why was everyone turning where I was turning? Hurry up. Hurry the &%$# up, people.

I turned into the school driveway and was quickly sandwiched between school buses and the entire Springton Middle School parent population in line for early (snow-day) pick-up. Ugh. I called Win. No answer. Expletive. I drove down a one-way drive; it might not have been a road. I had to get out of there and find Colt Road, the road originally mentioned in the text alert.

I drove right past Win and then recognized Jeff Gordon, then president of the American Birding Association. "Have you seen it?" Win yelled. "No," I responded frantically. "It's behind that house," and he gestured across the street. Among the larger Canada Geese stood our bird. Its orange legs jumped out at me.

Tundra Bean-Goose*, with orange legs (Media, Pennsylvania)*

Absolutely incredible. And I was on my 12:30 p.m. work call without a problem.

54. Vermont Winter

There is one thing as satisfying as chasing and seeing a new bird, and that is witnessing the joy in other birders as they add new birds to their life lists. The Christmas holiday week, family in Vermont, plus COVID combined to produce a wonderful socially distanced activity—birding! Brother-in-law Joe was already an avid birder, and now his son Carson was beginning his life list. On top of that, future son-in-law Hunter couldn't resist participating in the action.

I have spent countless evenings walking the Wilderness Community up there with limited success in hearing the resident Barred Owl. When Joe and Carson spotted one sitting in the apple tree outside of the Weisbord house at lunchtime, the text exchanges heated up and ultimately created plenty of excitement for everyone.

Barred Owl *(Mount Holly, Vermont)*
Photo by Carson Weisbord

Now that several family members were eagerly talking about birding and posting each sighting on eBird, I felt compelled to try to find them another special treat of Vermont's harsh winter climate. But what bird? A quick search of recent sightings in Rutland County showed that Evening Grosbeaks were being reported on a consistent basis only twelve miles away in Shrewsbury. These gorgeous birds are hefty members of the finch family that breed in Canada. Their appearance in the Northeast in winter is highly variable.

The four of us took off the next morning with the posted coordinates as our destination. Unfortunately, Waze led us up a dead-end road adjacent to private property. These coordinates were several miles from the written description "Vicky" had provided on her eBird posts. We tried the "intersection of Town Hill and Lincoln Hill Roads." Nothing. Not even a crow. We headed home a bit disappointed but determined to try again that afternoon.

Fueled by anticipation mixed with innate competitiveness, we returned to the promising intersection of the aforementioned roads. Two Eastern Bluebirds had caught our attention when we were politely interrupted. "Are you guys birders?" asked the driver of the only car we'd encountered in twenty minutes. I lowered my binoculars and asked: "Are you Vicky?" "Why yes," she replied and after a quick exchange of pleasantries, we learned that the birds she had been posting on eBird were actually seen in her backyard—a half mile from the intersection. She kindly allowed us to stand in the cover of trees and watch her feeders as the sun set. No grosbeaks, but several other new birds for Carson and Hunter elevated everyone's spirits. Of course, we invited ourselves back to Vicky's home the next morning and she enthusiastically obliged. Despite harsh winds and intermittent snow, more than ten different species gorged

Joe Weisbord, Carson Weisbord, and George (Mount Holly, Vermont)

themselves at her feeders. Seconds turned into minutes and after half an hour I was beginning to worry that we would miss our target bird. Then Hunter calmly asked, "What is that on the table?" Grosbeak! One, then three, ten, twenty! The bright yellow and black-and-white males lit up the feeding area.

2020 was a tough year in many ways, but this was a great way to close it.

Evening Grosbeaks *(Shrewsbury, Vermont)*
Photo by Vicky Arthur

55. Spring Break 2021

Fully vaccinated—thank you, Pfizer and Temple University—I boarded an airplane in March 2021 for the first time in thirteen months. Destination: Des Moines, Iowa. For the past few weeks, I had been tracking a Code-2 Smith's Longspur sighting in the Midwest. I have always known it would take a special trip to our country's heartland in the winter to have a decent chance of seeing this secretive LBJ.

A month earlier I had participated in a friend's podcast in which TC Scornavacchi interviewed me about my passion for birds.[6] After our humorous exchange and discussion of birds, I realized now was the time to chase this sparrow-like bird whose camouflaged coloring blends in perfectly with Iowa's open prairie grassy areas. It prefers three-awn grasses, *Aristida purpurea*, which are disappearing rapidly due to development. Because the bird breeds in the Arctic tundra, my greatest chance to see one was as it prepared to migrate.

I had thought Oklahoma was my best bet. Then the Little Rock, Arkansas, airport. But after many consultations with eBird and correspondences with local ornithologists, Iowan Aaron Brees convinced me that a trip to the Chichaqua Bottoms-Swan Lake in Elkhart, Iowa, was very possibly going to yield me ABA Area lifer number 789.

After arriving at the exact location, I walked the area, and within minutes I heard the longspur's diagnostic ticking rattle. These elusive six-inch ground-dwelling birds give that call

6 See TC Scornavacchi's tcafterdarkpodcast.com

right before alighting and quickly disappearing into the sky—pretty much before one can put a binocular on them. I paced back and forth over the several acres, occasionally disturbing and hearing a Smith's. But as I've stated more than once, I don't count heard-only birds. I want to see it and photograph it.

Finally, after two and a half hours of slogging through mud and manure, I detected a movement on the ground within range of my camera. Yes. Not only did I finally have a long enough look through my bins to confirm this was a Smith's, as opposed to its much more common relative the Lapland Longspur, I was able to capture a couple of images.

Smith's Longspur *(Elkhart, Iowa)*

What does one do when they find their target bird in Iowa, with time to spare? Drive six hours west to Kearney, Nebraska, the Sandhill Crane capital and one of the oldest and largest migrations in the world. More than 600,000 of these majestic birds—80 percent of the world's crane population—congregate along an eighty-mile stretch of the Platte River in March to fatten up on waste grain in the empty cornfields in preparation for the journey to their Arctic breeding grounds. What were the chances I was going to find the Code-4 Common Crane, a Eurasian species, that had been reported here two days earlier?

With the assistance of my new birding friend, Dave Cunningham, my chances were greatly improved. Dave was connected to a couple of hotshot Nebraskan birders, Doug and Bonnie. Despite the odds, Doug was able to locate the bird in a farmer's field among thousands of similarly sized Sandhills. Doug alerted Dave, who called me.

When I arrived at the spot, the others let me have scope views. I could identify the bird, barely. But what about a photograph? My digiscoping skills are marginal. Truth be told, they are non-existent. I have been in this position before. I suggested that they keep an eye on the bird while I try to meet the landowner and garner permission to walk in his fields for a better view and photograph.

Wyatt turned out to be a lovely fellow. I called Dave, and five minutes later the gang descended on Wyatt's property. Wyatt's eight-year-old son, Lucas, led the way on his ATV to a much closer viewing spot.

Common Crane *(Gibbon, Nebraska)*

The black face and neck are the Common Crane's distinguishing field marks. Lifer number 790. Thank you to my new Nebraskan birding buddies. I think I'll have a beer.

An ABA Area first-ever (Cuban subspecies) Code-5 Red-legged Thrush had been sporadically seen in the Key West Tropical Forest & Botanical Garden since October 2020. Because of COVID, I really had not thought about traveling to try to see this bird. Besides, it wasn't going to stick around this densely forested fifteen-acre preserve for very long anyway. Wrong! I was surprised to discover on eBird that this strikingly beautiful thrush was still present as late as March 26.

Thanks to Southwest Airlines, I was able to leave 40°F Omaha and the land of the cranes in the morning and search for the thrush in 85°F that same evening. No luck, though. The fact that it barely vocalizes makes it very difficult to locate this member of the *Turdidae* family. It's a relative of our American Robin.

Six hours in the tropical forest the following day, accompanied by birders from California, New Hampshire, and Ohio, yielded some interesting iguana and migrant-warbler sightings but no thrush for us.

Earlier that day, Cornell's Chris Wood, the creator of eBird, had briefly seen and photographed our bird, so we knew it was still present. Oh well, at least I was able to have dinner with my Miami-based daughter, Sarah.

A quick change in travel plans allowed me to spend an extra day looking for this reticent native of the Caribbean. Miami isn't exactly next door to Key West, but Chris's eBird post fueled me. One of the California guys, a fellow with the third largest life list in America, saw the fruit-eater associating with catbirds for a split second near the butterfly garden.

Then it was gone. Three more hours elapsed, and my new birding acquaintances all departed for various reasons.

It was after 12:00 p.m., and I wondered how I was going to locate this silent bird on my own before the 4:00 p.m. closing time. I decided to stay around the butterfly garden. A pair of cardinals caught my eye, and as I raised my binoculars to check for a nest, I saw a dark shape move high in the tree behind them. I moved my eyepiece to the dark shape and a white throat jumped out at me. It was the Red-legged Thrush.

I was able to make it back to Miami in time to have another dinner with Sarah. This had been an incredibly satisfying, fun-filled spring break. And it was only Wednesday.

Red-legged Thrush *(Key West, Florida)*

56. Nevada and Idaho

I believe the most difficult bird to locate in the United States is the Himalayan Snowcock. Native to Pakistan, it was introduced to the Ruby Mountains in 1963 by the Nevada Game Commission to promote hunting and tourism. It is now considered wild, as an established population of fewer than 1,000 of these pheasant-sized birds exist above tree line at elevations greater than 11,000 feet.

The only way to see one is to helicopter through this rugged area—you may recall the harrowing scene in *The Big Year* movie—or hike to a location where a spotting scope is needed to scan the distant ridges.

A 3:00 a.m. departure from our hotel followed by a forty-five-minute drive and ninety-minute ascent from 7,000 to 9,000 feet put us at the only spot in America where one has a chance to view these creatures. They tend to be most active at dawn as they descend from the remote ridges to lower-level meadows to feed on plants and insects.

Within minutes of arriving at the designated viewing spot, a leader from another bird group exclaimed there were three snowcocks at the tippy top of the ridge. Too distant for binocular viewing, the camouflaged birds scurried in and out of view among crevices of the craggy rock face wall while, thankfully, everyone managed good looks through scopes. Number 792.

Himalayan Snowcock *(Ruby Mountains, Nevada)*

The Himalayan Snowcock sightings in Nevada were certainly equivalent to winning a birding medal, possibly gold, if such a thing existed. As you know, by that point, it was hard for me to add a new species to my ABA life list. What were the odds that I could add a new species to my total the same day as the Himalayan Snowcock adventure?

The Red Crossbill population in southern Idaho's Cassia County had recently been declared a separate species, and the ABA had added Cassia Crossbill to its checklist.

This thick-billed population is relatively sedentary (non-migratory) and thrives because of the absence of the red squirrel, a competitor for its primary food source, lodgepole pinecones. Wildside Nature Tours conveniently arranges a weekend when a skilled guide (Chris Brown) helps birders attempt to see both rarities.

Since we were able to add the snowcock so quickly, we had time to drive from Nevada's Ruby Mountains to a campground in the Sawtooth National Forest in Idaho's South Hills and search for the Cassia.

A couple of very cooperative Cassia Crossbills were feeding high up in the campground's lodgepole pines, exactly as the field guide states. Their large, hooked, crisscrossed bills enable them to efficiently crack the large cones and eat the nutritious seeds.

Full disclosure: my photographs were somewhat unimpressive. However, my new birding friend gave me permission to publish hers. Same bird, but her camera.

Map from iBird Plus app

Cassia Crossbill *(Sawtooth National Forest, Idaho)*
Photo by Kisa Weeman

57. Gambell (again)

You will recall that I had visited Nome and Gambell in May 2013 and picked up twenty-four lifers. More importantly, I met Aaron Lang of Wilderness Birding Adventures. As I approached the magical number of 800, the most time- and cost-efficient trip to close my gap (eight species) was a September trip. Aaron and Steve Heinl were the perfect guides for the nine of us. Talented birders and cooks with a great sense of humor, they made ten days in a crowded one-bathroom house quite pleasurable.

Little Bunting *(Gambell, Alaska)*

Why do birds bother to stay in Gambell? The fatigued travelers are attracted to the unique green wormwood vegetation and insects that exist around several boneyards. You may recall that these pits contain discarded whale and walrus bones that have been excavated for their ivory. The ivory is carved and sold by entrepreneurial Yupik natives.

I did flush and spot a different-looking bird on the third day. The group surrounded the area where I thought the bird had landed. Aaron was on it. "Little Bunting," he yelled. A Code-4 from Asia. And a significant enough find to make the NARBA airwaves that evening.

I felt pretty good, because I was in serious birding company and wanted to contribute something other than calling out a common Snow Bunting or an inaccurate identification.

Many of the migrants are small, indistinct, and shy. Photography is key to identifying them. Thankfully, I was able to add three members of the *Phylloscopidae* family (leaf warblers) to my life list, a page in the bird guides that I never paid much attention to because of the unlikely chance I would ever see and identify these plain birds.

The leaf warblers page from the National Geographic Field Guide to the Birds of North America

Each morning we trudged a half-mile through pea gravel to the end of the island so that by dawn we stood watch over the open water looking for seabirds. One morning it did not rain and we could see Russia thirty-five miles away. Remember Sarah Palin? Anyway, this group did not discuss politics, only birds. And I picked up Spectacled Eider and Kittlitz's Murrelet. I had expected three lifers, hoped for five, and now had six, bringing my total to 798. And I still had half of the trip remaining.

This next bird is so special that no matter the number for me, it deserved its own blogpost. I was standing in one of the boneyards, not far from our guide, when Aaron's walkie-talkie crackled inaudibly. The leader from another group was trying to alert us to something unusual on the shore of Gambell's Troutman Lake. Aaron asked for a repeated message, and "Middendorff's" came through clearly. His eyes widened, and he beckoned our scattered group to hustle over. Admittedly, I was not familiar with this one. I thought Middendorff's might be a sparrow.

Arctic Warbler *(Gambell, Alaska)*

Common Chiffchaff *(Gambell, Alaska)*

Aaron had hitched a trailer to his ATV for moments like this. Walking was not an option, as the bird was two miles away and not likely to hang around. We piled on and raced down the airport runway toward the distant assembled group.

While riding his ATV along the shoreline, their leader had miraculously stumbled upon the very secretive, Code-4 Middendorff's Grasshopper-Warbler. Endemic to Asia, these birds prefer to remain concealed in dense vegetation.

When we arrived, one of our gang politely asked if this chunky vagrant could slip away through the grass bordering the shoreline without our noticing. Their other leader responded emphatically, "No! It's as big as a (expletive) eagle." An exaggeration, of course, but he was right. And we photographers were astounded and pleased.

Middendorff's Grasshopper-Warbler *(Gambell, Alaska)*

58. Number 800!

After the excitement of Mr. Middendorff, my mind immediately shifted to the thought that number 800 had to be special. Really special. But I could not control things out there in Gambell, so I told myself to relax. Yeah, right. I couldn't sleep that night as I imagined the possibilities for number 800.

The Gray-tailed Tattler (Code-3) was statistically the most likely but not what I envisioned. Lesser Sand-Plover? Maybe. Another possibility was returning home stuck at 799. Not a bad outcome. But I was there to rack up new birds, and I didn't want to go "0-for" the last four days.

The day after Mr. Middendorff's appearance started like every other Gambell day. Coffee, two hours at the sea-watch, followed by a sweep of the Far and Circular Boneyards. The elusive Stonechat was on our minds as several of us climbed rocky slopes looking for that specialty. Nada. After lunch we kept to the routine of marching through the Near Boneyard when the call came. Far Boneyard, *Siberian Accentor*. What a cool name, I thought. I knew only a little about this Code-4. A native of northern Siberia, it's a small, shy, sparrow-like bird with beautifully streaked patterns on its head.

Aaron shuttled most of the participants via ATV while two of us power-walked the three-quarters of a mile at a very brisk pace. The entire group viewed this migrant up close.

S-A-T-I-S-F-A-C-T-I-O-N.

***Siberian Accentor**: number 800! (Gambell, Alaska)*

George and guide extraordinaire Aaron Lang

Truth be told, we found that tattler two days later.
Now what?

Gray-tailed Tattler *(Gambell, Alaska)*

Epilogue

Bird of the Century

The intrepid Steller's Sea-Eagle is basically an endangered species native to Russia and Japan. It had *never* been spotted in Canada or the Lower 48 states until the summer of 2021. Startingly, one was photographed in New Brunswick, Canada, in July. COVID and location made it impossible for anyone but the game wardens to see it. This vagrant disappeared after a few days, and a month later it showed up near Halifax, Nova Scotia.

After another brief visit, it wasn't located again until early December, in Taunton, Massachusetts. Again, a two-day stay and poof, the largest eagle in the world (twenty pounds with an eight-foot wingspan) was gone. Not to be found again until 3:30 p.m. on December 30, in Georgetown, Maine. National publications were tracking this incredible individual's travels.

I was only 268 miles from Georgetown on Friday, December 31. But it was New Year's Eve. Family and friends were planning to gather at our Vermont home. Was there time to chase this raptor, maintain relationships, and show up at work on Monday? Yes. Provided one sleeps only three hours and has a very understanding spouse, flexible houseguests, and a forgiving daughter and agreeable paddle tennis teammates who had already been calendared on January 1 and 2 for activities in Haverford.

Jeff Dingle, George, and Joe Knowles

My adventurous buddy, Joe Knowles, and I left Vermont at 4:00 a.m. on New Year's Day. We collected my lifelong friend Jeff Dingle, who had introduced me to birding in 1968, in Portsmouth, New Hampshire, and headed to Maine. By 9:00 a.m. we had not received any sighting reports, so we drove to the spot our bird had been seen a day earlier.

But where were the birders? We expected to see a crowd given that 250 people had posted on eBird December 31. Then Joe checked his phone for any possible listserv updates. The eagle could be seen from the Five Islands Lobster Company wharf.

We drove the foggy and winding 2.9 miles, in what felt like an hour, to the Five Islands Lobster Company located at the end of a pier. As we approached the seacoast, we saw dozens of cars lining the road. Our hearts raced as we knew the distant bird could disappear behind an island at any moment or become obscured by the accumulating fog.

Thankfully, the birding gods delivered an impossibly wonderful New Year's gift.

Steller's Sea-Eagle *(Georgetown, Maine)*

Code-5 Hat Trick

You may recall that the American Birding Association defines a Code-5 species as one that has been seen five or fewer times in the ABA Area, ever. I had seen fewer than ten Code-5 birds in my lifetime. When the opportunity to see three Code-5 birds in one weekend presented itself, there was no question I would try.

Logistically, frequent-flier miles pretty much covered the airfare and car rental for a South Texas swing combined with a slight side trip to Carlsbad, New Mexico. With an upcoming holiday weekend, seventy-two hours was sufficient time to have a chance to go three for three—a "hat trick" in sports parlance. Seeing one of these three species would be nice, two was my expectation, and three was simply too exciting to imagine.

The Central American Bat Falcon had *never* been reported in the United States until December 18, 2021. It is not even mentioned in the National Geographic field guide nor in the iBird app. Guaranteed to be coded a 5 when accepted by the ABA Records Committee, the Bat Falcon was being seen fairly regularly in the Santa Ana National Wildlife Refuge near McAllen, Texas. It had developed a pattern of perching on a certain telephone pole on Highway 281 outside the refuge at crepuscular hours.

My flight from Houston to McAllen arrived in time for me to wait and watch with sixty other birders on a foggy Thursday evening. Nothing. A similarly sized group assembled ten hours later outside the refuge entrance before Friday's dawn. A flyby American Kestrel elicited several gasps from the anxious crowd. The kestrel is a relative of and similar in appearance to our Batman-looking falcon. Three hours later we had yet to see it. At this point, I decided to drive the sixty miles to the University of Texas Rio Grande Valley (UTRGV) Brownsville campus and chase the Code-5 Social Flycatcher, then return for another crack at the Bat Falcon.

The Social Flycatcher had been hanging around UTRGV for a couple of weeks. Once I figured out that I had been searching for an hour from the wrong bridge on campus, I found several birders focused on branches overhanging a body of water. I heard the diagnostic "squeaky toy" call, and bam—there was the Social in my view.

Social Flycatcher *(Brownsville, Texas)*

After picking up the Social, I birded my way back along the Rio Grande Valley, returning to the Bat Falcon stakeout pole at 4:00 p.m. Patience is a virtue, and birdwatching has helped me develop this weakness of mine. More than an hour passed and nothing except some birdy conversations with folks from Illinois, California, and Corpus Christi had ensued. Then, at 5:50 p.m., a fellow far to my left hollered: "That's it! Flying toward the pole." For the next twenty minutes the Bat Falcon posed on the telephone pole in fading light.

I was so full of excitement from the success of this last-minute appearance that the 6:00 a.m. flight to El Paso and the subsequent two-hour drive to pursue the elusive Code-5 Blue Mockingbird in New Mexico were not an inconvenience. eBird was providing hourly reports, and the chances of seeing it Saturday afternoon were fifty-fifty. Upon arriving at the much-publicized Rattlesnake Campground spot at 12:45 p.m., I was surprised that only one other birder was scouting the vast area.

Bat Falcon *(McAllen, Texas)*

I patiently waited near the deer skull marking a creekside cutleaf hackberry bush where the shy bird had been seen at 7:00 a.m. One hour, then two, then three hours passed. Several birders came and went.

I ticked eighteen species, including a Hermit Thrush and a Brown Thrasher in the distant copse across the stream. Both looked very much like our target at first glimpse. But no. At 5:00 p.m. the sun had dropped below the tree line, and I was beginning to think about reserving a hotel in Carlsbad and returning Sunday morning. My flight home was booked for Monday, anyway.

The quiet was interrupted by a mockingbird-ish sound to my right. I stood up from my log seat and walked around the corner. Just maybe. Please. Three guys were viewing the interior of a nearby fruiting hackberry tree. Blue Mockingbird. It disappeared across the creek after ninety seconds.

Blue Mockingbird *(Carlsbad, New Mexico)*

Palo Alto

The Oriental Turtle-Dove ranges mainly from Asia to Japan. It has been seen three times in California, but realistically the chances of seeing one in the United States are basically 0 percent.

I received a NARBA text that a presumed wild Oriental Turtle-Dove was identified in Palo Alto on February 2, 2022, while I was preparing for our February 10 alumni gathering in San Francisco. Frankly, I had never heard of this species. It doesn't even have a range map in the field guides, but heck, a potential lifer is always exciting. From February 2 through February 10 the bird was continuously seen for a couple of minutes every morning around 7:30 a.m. before disappearing into a residential area. I arrived at dawn on Friday, February 11, and took my place among fifteen or so anxious birders at the stakeout area.

By 7:30 a.m. the crowd had grown to over fifty. One elderly lady mumbled, "It isn't going to show" and walked despondently back to her car. I wondered how she could give up so easily, but maybe she was a local and had already seen it. By 8:15 a.m. I thought she could be right. Then at 8:25 a.m. a voice rang out, "In the redwood!" It was right in front of me, about fifty yards away. Binoculars first for the positive ID and then quick, camera—need a photo for my blog. No more than thirty seconds elapsed, and it was gone. But I knew I had captured it. I had a confident feeling of exhilaration similar to Chloe Kim, the Olympic snowboarder who nailed her first half-pipe and won gold. A reach comparison, perhaps, but not for me. Check out the diagnostic orange-red eye.

Oriental Turtle-Dove *(Palo Alto, California)*

During the chaotic moment when birders were scrambling to view and photograph the dove, I noticed a young guy who I thought could be Miles Horton, the son of our friends I'd shown the rare Garganey. I am very proud to write that he caught the birding bug after that special experience. But by the time I had snapped my photos, calmed down and gathered my wits, he was gone. I texted him, and immediately my phone rang. He and his friend had indeed witnessed the same spectacle I had. The pleasure of seeing them and the satisfaction I felt knowing I had helped inspire a lifelong hobby in another person far outweighed seeing ABA Area number 807.

Stewarding Mother Earth

To give you more perspective on the numbers, there are one hundred or so living people in the world who report having topped 800 for the ABA Area. I am motivated by the numbers, of course, but more importantly, I believe that if I share my passion with people like you, our world will be a more informed and better place. Climate change is real. Bird species are vanishing at a disturbing rate. And that means other animals are in peril, as well.

I show this photo of a polar bear I took fifteen years ago in Nunavut, Canada, because as much as I care about birds, I care mostly about nature. The polar bear, whose plight is well documented, is a more widely recognized symbol of our wild, untamed natural world. If my young grandchild Areli Bereolos is going to see a polar bear in the wild, we must do a better job of stewarding Mother Earth.

Granddaughter Areli Bereolos (Haverford, Pennsylvania)

APPENDICES

Appendix One: ABA Area Life List

ABA FIELD CHECKLIST

American Birding®
ASSOCIATION

Date 1968 - PRESENT Total Species 807
Locality USA + CANADA (ABA AREA)
Notes LIFETIME CHECKLIST (EX - MEGA RARITIES)

Taxonomy and nomenclature follow the 7th edition (1998) and all subsequent annual supplements of the American Ornithologist's Union Check-list (checklist.aou.org/taxa). Changes to the ABA Checklist are updated at aba.org/checklist and explained in Birding and Birder's Guide.

For general information on the ABA, to join the ABA, or to order additional copies of the ABA Field Checklist, please call 800-850-2473 or visit us online at aba.org.

Ducks, Geese, and Swans (Anatidae)

■ Black-bellied Whistling-Duck
■ Fulvous Whistling-Duck
■ Emperor Goose
■ Snow Goose
■ Ross's Goose
■ Greater White-fronted Goose
❑ Taiga Bean-Goose
■ Tundra Bean-Goose
■ Brant
■ Cackling Goose
■ Canada Goose
■ Hawaiian Goose
■ Mute Swan
■ Trumpeter Swan
■ Tundra Swan
❑ Whooper Swan
■ Egyptian Goose
■ Muscovy Duck
■ Wood Duck
■ Blue-winged Teal
■ Cinnamon Teal
■ Northern Shoveler
■ Gadwall
■ Eurasian Wigeon
■ American Wigeon
❑ Laysan Duck
■ Hawaiian Duck
■ Mallard
■ American Black Duck
■ Mottled Duck
■ Northern Pintail
■ Green-winged Teal
■ Canvasback
■ Redhead
❑ Common Pochard
■ Ring-necked Duck
■ Tufted Duck
■ Greater Scaup
■ Lesser Scaup
■ Steller's Eider
■ Spectacled Eider
■ King Eider
■ Common Eider
■ Harlequin Duck
■ Surf Scoter
■ White-winged Scoter
■ Black Scoter
■ Long-tailed Duck
■ Bufflehead
■ Common Goldeneye
■ Barrow's Goldeneye
❑ Smew
■ Hooded Merganser
■ Common Merganser
■ Red-breasted Merganser
❑ Masked Duck
■ Ruddy Duck

Curassows and Guans (Cracidae)

■ Plain Chachalaca

New World Quail (Odontophoridae)

■ Mountain Quail
■ Northern Bobwhite
■ Scaled Quail
■ California Quail
■ Gambel's Quail
■ Montezuma Quail

Partridges, Grouse, Turkeys, and Old World Quail (Phasianidae)

■ Chukar
■ Gray Francolin
■ Black Francolin
■ Erckel's Francolin
■ Himalayan Snowcock
■ Gray Partridge
■ Red Junglefowl
■ Kalij Pheasant
■ Ring-necked Pheasant
■ Ruffed Grouse
■ Greater Sage-Grouse
■ Gunnison Sage-Grouse
■ Spruce Grouse
■ Willow Ptarmigan
■ Rock Ptarmigan
■ White-tailed Ptarmigan
■ Dusky Grouse
■ Sooty Grouse
■ Sharp-tailed Grouse
■ Greater Prairie-Chicken
■ Lesser Prairie-Chicken
■ Wild Turkey

Flamingos (Phoenicopteridae)

■ American Flamingo

Grebes (Podicipedidae)

■ Least Grebe

- [x] Pied-billed Grebe
- [x] Horned Grebe
- [x] Red-necked Grebe
- [x] Eared Grebe
- [x] Western Grebe
- [x] Clark's Grebe

Sandgrouses (Pteroclidae)

- [ ] Chestnut-bellied Sandgrouse

Pigeons and Doves (Columbidae)

- [x] Rock Pigeon
- [x] White-crowned Pigeon
- [x] Red-billed Pigeon
- [x] Band-tailed Pigeon
- [x] Eurasian Collared-Dove
- [x] Spotted Dove
- [x] Zebra Dove
- [x] Inca Dove
- [x] Common Ground-Dove
- [x] Ruddy Ground-Dove
- [x] White-tipped Dove
- [x] White-winged Dove
- [x] Mourning Dove

Cuckoos, Roadrunners, and Anis (Cuculidae)

- [x] Common Cuckoo
- [x] Yellow-billed Cuckoo
- [x] Mangrove Cuckoo
- [x] Black-billed Cuckoo
- [x] Greater Roadrunner
- [x] Smooth-billed Ani
- [x] Groove-billed Ani

Goatsuckers (Caprimulgidae)

- [x] Lesser Nighthawk
- [x] Common Nighthawk
- [ ] Antillean Nighthawk
- [x] Common Pauraque
- [x] Common Poorwill
- [x] Chuck-will's-widow
- [x] Buff-collared Nightjar
- [x] Eastern Whip-poor-will
- [x] Mexican Whip-poor-will

Swifts (Apodidae)

- [ ] Black Swift
- [ ] White-collared Swift
- [x] Chimney Swift
- [x] Vaux's Swift
- [x] Mariana Swiftlet
- [x] White-throated Swift

Hummingbirds (Trochilidae)

- [ ] Mexican Violetear
- [ ] Rivoli's Hummingbird
- [x] Blue-throated Hummingbird
- [x] Lucifer Hummingbird
- [x] Ruby-throated Hummingbird
- [x] Black-chinned Hummingbird
- [x] Anna's Hummingbird
- [x] Costa's Hummingbird
- [x] Broad-tailed Hummingbird
- [x] Rufous Hummingbird
- [x] Allen's Hummingbird
- [x] Calliope Hummingbird
- [x] Broad-billed Hummingbird
- [ ] Berylline Hummingbird
- [x] Buff-bellied Hummingbird
- [x] Violet-crowned Hummingbird
- [x] White-eared Hummingbird

Rails, Gallinules, and Coots (Rallidae)

- [ ] Yellow Rail
- [x] Black Rail
- [x] Ridgway's Rail
- [x] Clapper Rail
- [x] King Rail
- [x] Virginia Rail
- [x] Sora
- [x] Purple Gallinule
- [x] Purple Swamphen
- [x] Common Gallinule
- [ ] Common Moorhen
- [ ] Eurasian Coot
- [x] Hawaiian Coot
- [x] American Coot

Limpkins (Aramidae)

- [x] Limpkin

Cranes (Gruidae)

- [x] Sandhill Crane
- [x] Whooping Crane

Stilts and Avocets (Recurvirostridae)

- [x] Black-necked Stilt
- [x] American Avocet

Oystercatchers (Haematopodidae)

- [ ] Eurasian Oystercatcher
- [x] American Oystercatcher
- [x] Black Oystercatcher

Lapwings and Plovers (Charadriidae)

- [x] Black-bellied Plover
- [x] American Golden-Plover
- [x] Pacific Golden-Plover
- [ ] Lesser Sand-Plover
- [x] Snowy Plover
- [x] Wilson's Plover
- [x] Common Ringed Plover
- [x] Semipalmated Plover
- [x] Piping Plover
- [x] Killdeer
- [x] Mountain Plover

Sandpipers, Phalaropes, and Allies (Scolopacidae)

- [x] Upland Sandpiper
- [x] Bristle-thighed Curlew
- [x] Whimbrel
- [x] Long-billed Curlew
- [x] Bar-tailed Godwit
- [x] Black-tailed Godwit
- [x] Hudsonian Godwit
- [x] Marbled Godwit
- [x] Ruddy Turnstone
- [x] Black Turnstone
- [x] Red Knot
- [x] Surfbird
- [x] Ruff
- [x] Sharp-tailed Sandpiper
- [x] Stilt Sandpiper
- [x] Curlew Sandpiper
- [ ] Temminck's Stint
- [ ] Long-toed Stint
- [x] Red-necked Stint
- [x] Sanderling
- [x] Dunlin
- [x] Rock Sandpiper
- [x] Purple Sandpiper
- [x] Baird's Sandpiper
- [x] Least Sandpiper
- [x] White-rumped Sandpiper
- [x] Buff-breasted Sandpiper
- [x] Pectoral Sandpiper
- [x] Semipalmated Sandpiper
- [x] Western Sandpiper
- [x] Short-billed Dowitcher
- [x] Long-billed Dowitcher
- [x] American Woodcock
- [ ] Common Snipe
- [x] Wilson's Snipe
- [ ] Terek Sandpiper
- [ ] Common Sandpiper
- [x] Spotted Sandpiper
- [x] Solitary Sandpiper
- [x] Gray-tailed Tattler
- [x] Wandering Tattler
- [x] Lesser Yellowlegs
- [x] Willet
- [x] Common Greenshank
- [x] Greater Yellowlegs
- [ ] Wood Sandpiper
- [x] Wilson's Phalarope
- [x] Red-necked Phalarope
- [x] Red Phalarope

Skuas and Jaegers (Stercorariidae)

- [ ] Great Skua
- [x] South Polar Skua
- [x] Pomarine Jaeger
- [x] Parasitic Jaeger
- [x] Long-tailed Jaeger

Auks, Murres, and Puffins (Alcidae)

- [x] Dovekie
- [x] Common Murre
- [x] Thick-billed Murre
- [x] Razorbill
- [x] Black Guillemot
- [x] Pigeon Guillemot
- [ ] Long-billed Murrelet
- [x] Marbled Murrelet

■ Kittlitz's Murrelet
■ Scripps's Murrelet
❑ Guadalupe Murrelet
❑ Craveri's Murrelet
■ Ancient Murrelet
■ Cassin's Auklet
■ Parakeet Auklet
■ Least Auklet
❑ Whiskered Auklet
■ Crested Auklet
■ Rhinoceros Auklet
■ Atlantic Puffin
■ Horned Puffin
■ Tufted Puffin

Gulls, Terns, and Skimmers (Laridae)

■ Black-legged Kittiwake
■ Red-legged Kittiwake
■ Ivory Gull
■ Sabine's Gull
■ Bonaparte's Gull
■ Black-headed Gull
■ Little Gull
■ Ross's Gull
■ Laughing Gull
■ Franklin's Gull
■ Heermann's Gull
■ Mew Gull
■ Ring-billed Gull
■ Western Gull
■ Yellow-footed Gull
■ California Gull
■ Herring Gull
■ Iceland Gull
■ Lesser Black-backed Gull
■ Slaty-backed Gull
■ Glaucous-winged Gull
■ Glaucous Gull
■ Great Black-backed Gull
■ Brown Noddy
■ Black Noddy
❑ Blue-gray Noddy
■ White Tern
■ Sooty Tern
❑ Gray-backed Tern
■ Bridled Tern
■ Aleutian Tern
■ Least Tern
■ Gull-billed Tern
■ Caspian Tern
■ Black Tern
■ Roseate Tern
■ Common Tern
■ Arctic Tern
■ Forster's Tern
■ Royal Tern
■ Sandwich Tern
■ Elegant Tern
■ Black Skimmer

Tropicbirds (Phaethontidae)

■ White-tailed Tropicbird
❑ Red-billed Tropicbird
■ Red-tailed Tropicbird

Loons (Gaviidae)

■ Red-throated Loon
■ Arctic Loon
■ Pacific Loon
■ Common Loon
■ Yellow-billed Loon

Albatrosses (Diomedeidae)

■ Laysan Albatross
■ Black-footed Albatross
❑ Short-tailed Albatross

Shearwaters and Petrels (Procellariidae)

■ Northern Fulmar
■ Trindade Petrel
■ Murphy's Petrel
❑ Mottled Petrel
❑ Bermuda Petrel
■ Black-capped Petrel
❑ Juan Fernandez Petrel
■ Hawaiian Petrel
❑ White-necked Petrel
❑ Bonin Petrel
❑ Black-winged Petrel
❑ Fea's Petrel
❑ Cook's Petrel
❑ Bulwer's Petrel
■ Cory's Shearwater
■ Wedge-tailed Shearwater
■ Buller's Shearwater
■ Short-tailed Shearwater
■ Sooty Shearwater
■ Great Shearwater
■ Pink-footed Shearwater
■ Flesh-footed Shearwater
❑ Christmas Shearwater
■ Manx Shearwater
❑ Newell's Shearwater
■ Black-vented Shearwater
■ Audubon's Shearwater

Storm-Petrels (Hydrobatidae)

■ Wilson's Storm-Petrel
❑ White-faced Storm-Petrel
■ Fork-tailed Storm-Petrel
■ Leach's Storm-Petrel
❑ Townsend's Storm-Petrel
■ Ashy Storm-Petrel
■ Band-rumped Storm-Petrel
■ Black Storm-Petrel
❑ Tristram's Storm-Petrel
❑ Least Storm-Petrel

Storks (Ciconiidae)

■ Wood Stork

Frigatebirds (Fregatidae)

■ Magnificent Frigatebird
■ Great Frigatebird

Boobies and Gannets (Sulidae)

■ Masked Booby
■ Brown Booby
❑ Red-footed Booby
■ Northern Gannet

Cormorants (Phalacrocoracidae)

■ Brandt's Cormorant
■ Neotropic Cormorant
■ Double-crested Cormorant
■ Great Cormorant
■ Red-faced Cormorant
■ Pelagic Cormorant

Darters (Anhingidae)

■ Anhinga

Pelicans (Pelecanidae)

■ American White Pelican
■ Brown Pelican

Bitterns, Herons, and Allies (Ardeidae)

■ American Bittern
■ Least Bittern
■ Great Blue Heron
■ Great Egret
■ Snowy Egret
■ Little Blue Heron
■ Tricolored Heron
■ Reddish Egret
■ Cattle Egret
■ Green Heron
■ Black-crowned Night-Heron
■ Yellow-crowned Night-Heron

Ibises and Spoonbills (Threskiornithidae)

■ White Ibis
■ Glossy Ibis
■ White-faced Ibis
■ Roseate Spoonbill

New World Vultures (Cathartidae)

■ Black Vulture
■ Turkey Vulture
■ California Condor

Ospreys (Pandionidae)

■ Osprey

Hawks, Kites, Eagles, and Allies (Accipitridae)

■ Hook-billed Kite
■ Swallow-tailed Kite
■ White-tailed Kite
■ Snail Kite
■ Mississippi Kite
■ Bald Eagle
■ Northern Harrier
■ Sharp-shinned Hawk
■ Cooper's Hawk
■ Northern Goshawk
■ Common Black Hawk

Harris's Hawk
White-tailed Hawk
Gray Hawk
Red-shouldered Hawk
Broad-winged Hawk
Hawaiian Hawk
Short-tailed Hawk
Swainson's Hawk
Zone-tailed Hawk
Red-tailed Hawk
Rough-legged Hawk
Ferruginous Hawk
Golden Eagle

Barn Owls (Tytonidae)

Barn Owl

Typical Owls (Strigidae)

Flammulated Owl
Western Screech-Owl
Eastern Screech-Owl
Whiskered Screech-Owl
Great Horned Owl
Snowy Owl
Northern Hawk Owl
Northern Pygmy-Owl
Ferruginous Pygmy-Owl
Elf Owl
Burrowing Owl
Spotted Owl
Barred Owl
Great Gray Owl
Long-eared Owl
Short-eared Owl
Boreal Owl
Northern Saw-whet Owl

Trogons (Trogonidae)

Elegant Trogon
Eared Quetzal

Kingfishers (Alcedinidae)

Ringed Kingfisher
Belted Kingfisher
Green Kingfisher

Woodpeckers and Allies (Picidae)

Lewis's Woodpecker
Red-headed Woodpecker
Acorn Woodpecker
Gila Woodpecker
Golden-fronted Woodpecker
Red-bellied Woodpecker
Williamson's Sapsucker
Yellow-bellied Sapsucker
Red-naped Sapsucker
Red-breasted Sapsucker
Ladder-backed Woodpecker
Nuttall's Woodpecker
Downy Woodpecker
Hairy Woodpecker
Arizona Woodpecker
Red-cockaded Woodpecker
White-headed Woodpecker
American Three-toed Woodpecker
Black-backed Woodpecker
Northern Flicker
Gilded Flicker
Pileated Woodpecker

Caracaras and Falcons (Falconidae)

Crested Caracara
American Kestrel
Merlin
Aplomado Falcon
Gyrfalcon
Peregrine Falcon
Prairie Falcon

Parakeets, Macaws, and Parrots (Psittacidae)

Monk Parakeet
Nanday Parakeet
Green Parakeet
White-winged Parakeet
Red-crowned Parrot

Lories, Lovebirds, and Australasian Parrots (Psittaculidae)

Rose-ringed Parakeet
Rosy-faced Lovebird

Tyrant Flycatchers (Tyrannidae)

Northern Beardless-Tyrannulet
Olive-sided Flycatcher
Greater Pewee
Western Wood-Pewee
Eastern Wood-Pewee
Yellow-bellied Flycatcher
Acadian Flycatcher
Alder Flycatcher
Willow Flycatcher
Least Flycatcher
Hammond's Flycatcher
Gray Flycatcher
Dusky Flycatcher
Pacific-slope Flycatcher
Cordilleran Flycatcher
Buff-breasted Flycatcher
Black Phoebe
Eastern Phoebe
Say's Phoebe
Vermilion Flycatcher
Dusky-capped Flycatcher
Ash-throated Flycatcher
Great Crested Flycatcher
Brown-crested Flycatcher
La Sagra's Flycatcher
Great Kiskadee
Sulphur-bellied Flycatcher
Tropical Kingbird
Couch's Kingbird
Cassin's Kingbird
Thick-billed Kingbird
Western Kingbird
Eastern Kingbird
Gray Kingbird
Scissor-tailed Flycatcher
Fork-tailed Flycatcher

Becards, Tityras, and Allies (Tityridae)

Rose-throated Becard

Shrikes (Laniidae)

Loggerhead Shrike
Northern Shrike

Vireos (Vireonidae)

Black-capped Vireo
White-eyed Vireo
Bell's Vireo
Gray Vireo
Hutton's Vireo
Yellow-throated Vireo
Cassin's Vireo
Blue-headed Vireo
Plumbeous Vireo
Philadelphia Vireo
Warbling Vireo
Red-eyed Vireo
Yellow-green Vireo
Black-whiskered Vireo

Jays and Crows (Corvidae)

Gray Jay
Green Jay
Pinyon Jay
Steller's Jay
Blue Jay
Florida Scrub-Jay
Island Scrub-Jay
California Scrub-Jay
Woodhouse's Scrub-Jay
Mexican Jay
Clark's Nutcracker
Black-billed Magpie
Yellow-billed Magpie
American Crow
Northwestern Crow
Fish Crow
Chihuahuan Raven
Common Raven

Monarch Flycatchers (Monarchidae)

Kauai Elepaio
Oahu Elepaio
Hawaii Elepaio

Larks (Alaudidae)

Eurasian Skylark
Horned Lark

Swallows (Hirundinidae)

Purple Martin
Tree Swallow

■ Violet-green Swallow
■ Northern Rough-winged Swallow
■ Bank Swallow
■ Cliff Swallow
■ Cave Swallow
■ Barn Swallow

Chickadees and Titmice (Paridae)

■ Carolina Chickadee
■ Black-capped Chickadee
■ Mountain Chickadee
■ Mexican Chickadee
■ Chestnut-backed Chickadee
■ Boreal Chickadee
❑ Gray-headed Chickadee
■ Bridled Titmouse
■ Oak Titmouse
■ Juniper Titmouse
■ Tufted Titmouse
■ Black-crested Titmouse

Verdin (Remizidae)

■ Verdin

Bushtits (Aegithalidae)

■ Bushtit

Nuthatches (Sittidae)

■ Red-breasted Nuthatch
■ White-breasted Nuthatch
■ Pygmy Nuthatch
■ Brown-headed Nuthatch

Creepers (Certhiidae)

■ Brown Creeper

Wrens (Troglodytidae)

■ Rock Wren
■ Canyon Wren
■ House Wren
■ Pacific Wren
■ Winter Wren
■ Sedge Wren
■ Marsh Wren
■ Carolina Wren
■ Bewick's Wren
■ Cactus Wren

Gnatcatchers and Gnatwrens (Polioptilidae)

■ Blue-gray Gnatcatcher
■ California Gnatcatcher
■ Black-tailed Gnatcatcher
■ Black-capped Gnatcatcher

Dippers (Cinclidae)

■ American Dipper

Bulbuls (Pycnonotidae)

■ Red-vented Bulbul
■ Red-whiskered Bulbul

Kinglets (Regulidae)

■ Golden-crowned Kinglet
■ Ruby-crowned Kinglet

Bush-Warblers (Cettiidae)

■ Japanese Bush-Warbler

Leaf Warblers (Phylloscopidae)

■ Arctic Warbler

Sylviid Warblers (Sylviidae)

■ Wrentit

White-eyes (Zosteropidae)

■ Japanese White-eye

Laughingthrushes (Timaliidae)

❑ Greater Necklaced Laughingthrush
■ Hwamei
■ Red-billed Leiothrix

Reed Warblers (Acrocephalidae)

❑ Millerbird

Old World Flycatchers (Muscicapidae)

■ White-rumped Shama
❑ Siberian Rubythroat
■ Bluethroat
■ Northern Wheatear

Thrushes (Turdidae)

■ Eastern Bluebird
■ Western Bluebird
■ Mountain Bluebird
■ Townsend's Solitaire
■ Omao
❑ Puaiohi
■ Veery
■ Gray-cheeked Thrush
■ Bicknell's Thrush
■ Swainson's Thrush
■ Hermit Thrush
■ Wood Thrush
❑ Eyebrowed Thrush
■ Clay-colored Thrush
■ Rufous-backed Robin
■ American Robin
■ Varied Thrush

Mockingbirds and Thrashers (Mimidae)

■ Gray Catbird
■ Curve-billed Thrasher
■ Brown Thrasher
■ Long-billed Thrasher
■ Bendire's Thrasher
■ California Thrasher
■ LeConte's Thrasher
■ Crissal Thrasher
■ Sage Thrasher
■ Northern Mockingbird

Starlings (Sturnidae)

■ European Starling
■ Common Myna

Waxwings (Bombycillidae)

■ Bohemian Waxwing
■ Cedar Waxwing

Silky-flycatchers (Ptiliogonatidae)

■ Phainopepla

Olive Warblers (Peucedramidae)

■ Olive Warbler

Waxbills (Estrildidae)

■ Common Waxbill
❑ Red Avadavat
■ African Silverbill
■ Java Sparrow
■ Scaly-breasted Munia
■ Chestnut Munia

Old World Sparrows (Passeridae)

■ House Sparrow
■ Eurasian Tree Sparrow

Wagtails and Pipits (Motacillidae)

■ Eastern Yellow Wagtail
■ White Wagtail
❑ Olive-backed Pipit
■ Red-throated Pipit
■ American Pipit
■ Sprague's Pipit

Fringilline and Cardueline Finches and Allies (Fringillidae)

■ Brambling
■ Evening Grosbeak
❑ Akikiki
❑ Maui Alauahio
■ Palila
❑ Laysan Finch
❑ Nihoa Finch
❑ Akohekohe
■ Apapane
■ Iiwi
❑ Maui Parrotbill
■ Akiapolaau
■ Anianiau
■ Hawaii Amakihi
■ Oahu Amakihi
■ Kauai Amakihi
■ Hawaii Creeper
❑ Akekee
■ Hawaii Akepa
■ Pine Grosbeak
■ Gray-crowned Rosy-Finch
■ Black Rosy-Finch
■ Brown-capped Rosy-Finch
■ House Finch
■ Purple Finch
■ Cassin's Finch
■ Yellow-fronted Canary
■ Common Redpoll
■ Hoary Redpoll
■ Red Crossbill
■ Cassia Crossbill
■ White-winged Crossbill
■ Pine Siskin
■ Lesser Goldfinch

- [x] Lawrence's Goldfinch
- [x] American Goldfinch
- [ ] Island Canary

Longspurs and Snow Buntings (Calcariidae)

- [x] Lapland Longspur
- [x] Chestnut-collared Longspur
- [x] Smith's Longspur
- [x] McCown's Longspur
- [x] Snow Bunting
- [ ] McKay's Bunting

Emberizids (Emberizidae)

- [x] Rustic Bunting

Towhees and Sparrows (Passerellidae)

- [x] Olive Sparrow
- [x] Green-tailed Towhee
- [x] Spotted Towhee
- [x] Eastern Towhee
- [x] Rufous-crowned Sparrow
- [x] Canyon Towhee
- [x] California Towhee
- [x] Abert's Towhee
- [x] Rufous-winged Sparrow
- [x] Botteri's Sparrow
- [x] Cassin's Sparrow
- [x] Bachman's Sparrow
- [x] American Tree Sparrow
- [x] Chipping Sparrow
- [x] Clay-colored Sparrow
- [x] Brewer's Sparrow
- [x] Field Sparrow
- [x] Black-chinned Sparrow
- [x] Vesper Sparrow
- [x] Lark Sparrow
- [x] Five-striped Sparrow
- [x] Black-throated Sparrow
- [x] Sagebrush Sparrow
- [x] Bell's Sparrow
- [x] Lark Bunting
- [x] Savannah Sparrow
- [x] Grasshopper Sparrow
- [x] Baird's Sparrow
- [x] Henslow's Sparrow
- [x] LeConte's Sparrow
- [x] Nelson's Sparrow
- [x] Saltmarsh Sparrow
- [x] Seaside Sparrow
- [x] Fox Sparrow
- [x] Song Sparrow
- [x] Lincoln's Sparrow
- [x] Swamp Sparrow
- [x] White-throated Sparrow
- [x] Harris's Sparrow
- [x] White-crowned Sparrow
- [x] Golden-crowned Sparrow
- [x] Dark-eyed Junco
- [x] Yellow-eyed Junco

Spindalises (Spindalidae)

- [x] Western Spindalis

Yellow-breasted Chats (Icteriidae)

- [x] Yellow-breasted Chat

Blackbirds (Icteridae)

- [x] Yellow-headed Blackbird
- [x] Bobolink
- [x] Eastern Meadowlark
- [x] Western Meadowlark
- [x] Orchard Oriole
- [x] Hooded Oriole
- [x] Bullock's Oriole
- [x] Spot-breasted Oriole
- [x] Altamira Oriole
- [x] Audubon's Oriole
- [x] Baltimore Oriole
- [x] Scott's Oriole
- [x] Red-winged Blackbird
- [x] Tricolored Blackbird
- [x] Shiny Cowbird
- [x] Bronzed Cowbird
- [x] Brown-headed Cowbird
- [x] Rusty Blackbird
- [x] Brewer's Blackbird
- [x] Common Grackle
- [x] Boat-tailed Grackle
- [x] Great-tailed Grackle

Wood-Warblers (Parulidae)

- [x] Ovenbird
- [x] Worm-eating Warbler
- [x] Louisiana Waterthrush
- [x] Northern Waterthrush
- [x] Golden-winged Warbler
- [x] Blue-winged Warbler
- [x] Black-and-white Warbler
- [x] Prothonotary Warbler
- [x] Swainson's Warbler
- [x] Tennessee Warbler
- [x] Orange-crowned Warbler
- [x] Colima Warbler
- [x] Lucy's Warbler
- [x] Nashville Warbler
- [x] Virginia's Warbler
- [x] Connecticut Warbler
- [x] MacGillivray's Warbler
- [x] Mourning Warbler
- [x] Kentucky Warbler
- [x] Common Yellowthroat
- [x] Hooded Warbler
- [x] American Redstart
- [x] Kirtland's Warbler
- [x] Cape May Warbler
- [x] Cerulean Warbler
- [x] Northern Parula
- [x] Tropical Parula
- [x] Magnolia Warbler
- [x] Bay-breasted Warbler
- [x] Blackburnian Warbler
- [x] Yellow Warbler
- [x] Chestnut-sided Warbler
- [x] Blackpoll Warbler
- [x] Black-throated Blue Warbler
- [x] Palm Warbler
- [x] Pine Warbler
- [x] Yellow-rumped Warbler
- [x] Yellow-throated Warbler
- [x] Prairie Warbler
- [x] Grace's Warbler
- [x] Black-throated Gray Warbler
- [x] Townsend's Warbler
- [x] Hermit Warbler
- [x] Golden-cheeked Warbler
- [x] Black-throated Green Warbler
- [x] Rufous-capped Warbler
- [x] Canada Warbler
- [x] Wilson's Warbler
- [x] Red-faced Warbler
- [x] Painted Redstart

Cardinals, Piranga Tanagers and Allies (Cardinalidae)

- [x] Hepatic Tanager
- [x] Summer Tanager
- [x] Scarlet Tanager
- [x] Western Tanager
- [x] Flame-colored Tanager
- [x] Northern Cardinal
- [x] Pyrrhuloxia
- [x] Rose-breasted Grosbeak
- [x] Black-headed Grosbeak
- [x] Blue Grosbeak
- [x] Lazuli Bunting
- [x] Indigo Bunting
- [x] Varied Bunting
- [x] Painted Bunting
- [x] Dickcissel

Tanagers and Allies (Thraupidae)

- [x] Red-crested Cardinal
- [x] Yellow-billed Cardinal
- [x] Saffron Finch
- [ ] Yellow-faced Grassquit
- [x] White-collared Seedeater

Write-in

Appendix Two: 1969 South Florida Checklist

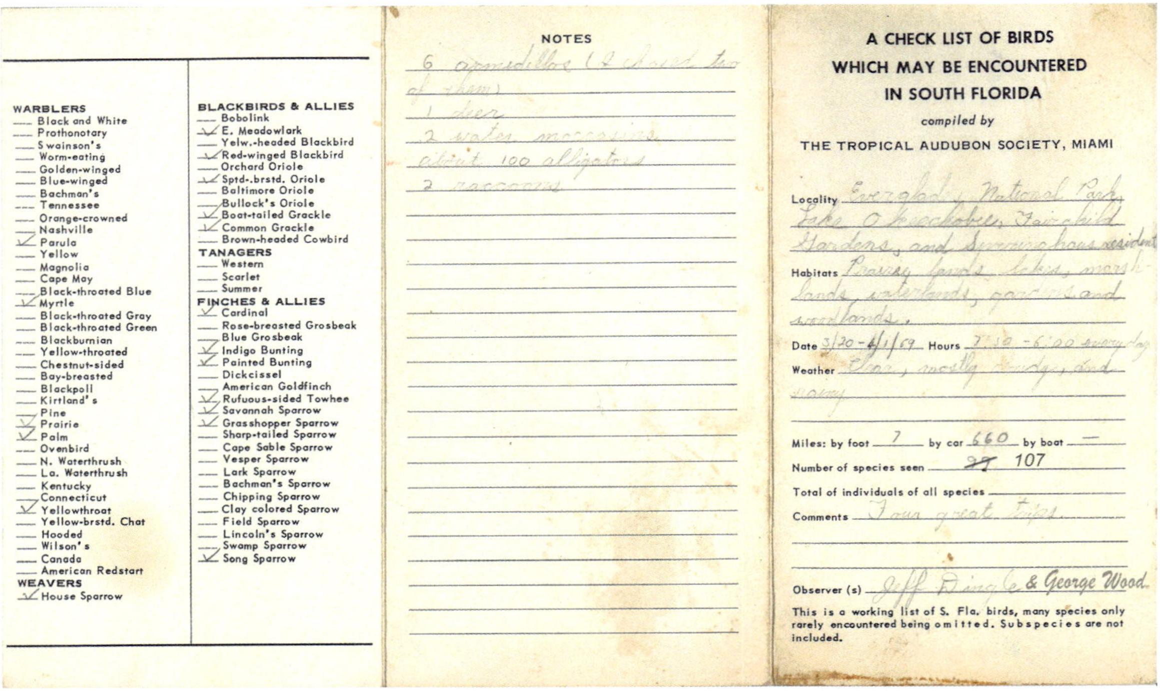
WARBLERS
- Black and White
- Prothonotary
- Swainson's
- Worm-eating
- Golden-winged
- Blue-winged
- Bachman's
- Tennessee
- Orange-crowned
- Nashville
- ✓ Parula
- Yellow
- Magnolia
- Cape May
- Black-throated Blue
- ✓ Myrtle
- Black-throated Gray
- Black-throated Green
- Blackburnian
- Yellow-throated
- Chestnut-sided
- Bay-breasted
- Blackpoll
- Kirtland's
- Pine
- ✓ Prairie
- ✓ Palm
- Ovenbird
- N. Waterthrush
- La. Waterthrush
- Kentucky
- Connecticut
- ✓ Yellowthroat
- Yellow-brstd. Chat
- Hooded
- Wilson's
- Canada
- American Redstart

WEAVERS
- ✓ House Sparrow

BLACKBIRDS & ALLIES
- Bobolink
- ✓ E. Meadowlark
- Yelw.-headed Blackbird
- ✓ Red-winged Blackbird
- Orchard Oriole
- ✓ Sptd-.brstd. Oriole
- Baltimore Oriole
- Bullock's Oriole
- ✓ Boat-tailed Grackle
- ✓ Common Grackle
- Brown-headed Cowbird

TANAGERS
- Western
- Scarlet
- Summer

FINCHES & ALLIES
- ✓ Cardinal
- Rose-breasted Grosbeak
- Blue Grosbeak
- ✓ Indigo Bunting
- ✓ Painted Bunting
- Dickcissel
- American Goldfinch
- ✓ Rufuous-sided Towhee
- ✓ Savannah Sparrow
- ✓ Grasshopper Sparrow
- Sharp-tailed Sparrow
- Cape Sable Sparrow
- Vesper Sparrow
- Lark Sparrow
- Bachman's Sparrow
- Chipping Sparrow
- Clay colored Sparrow
- Field Sparrow
- Lincoln's Sparrow
- Swamp Sparrow
- ✓ Song Sparrow

NOTES

6 armadillos (I chased two of them)
1 deer
2 water moccasins
about 100 alligators
2 raccoons

A CHECK LIST OF BIRDS
WHICH MAY BE ENCOUNTERED
IN SOUTH FLORIDA

compiled by
THE TROPICAL AUDUBON SOCIETY, MIAMI

Locality Everglades National Park, Lake Okeechobee, Fairchild Gardens, and Spring[illegible] house resident

Habitats Prairie lands, lakes, marsh lands, waterlands, gardens and woodlands.

Date 3/20-4/1/69 Hours 7:30-6:00 every day

Weather Clear, mostly cloudy, and rainy

Miles: by foot 7 by car 660 by boat —

Number of species seen ~~97~~ 107

Total of individuals of all species

Comments Four great trips.

Observer (s) Jeff Dingle & George Wood

This is a working list of S. Fla. birds, many species only rarely encountered being omitted. Subspecies are not included.

LOONS
- Common Loon
- Red-throated Loon

GREBES
- Horned Grebe
- ✓ Pied-billed Grebe

SHEARWATERS & PETRELS
- Greater Shearwater
- Sooty Shearwater
- Audubon's Shearwater
- Wilson's Petrel

PELICANS & ALLIES
- Wht.-tld. Tropic-bird
- ✓ White Pelican
- ✓ Brown Pelican
- Blue-faced Bobby
- Brown Booby
- Gannet
- ✓ Double-crest. Cormorant
- ✓ Anhinga
- ✓ Frigate-bird

HERONS & ALLIES
- ✓ Great White Heron
- ✓ Great Blue Heron
- ✓ Green Heron
- ✓ Little Blue Heron
- ✓ Cattle Egret
- ✓ Reddish Egret
- ✓ Common Egret
- ✓ Snowy Egret
- ✓ Louisiana Heron
- ✓ Blk.-cr. Night Heron
- ✓ Yelw.-cr. Night Heron
- Least Bittern
- ✓ American Bittern
- ✓ Wood Ibis
- ✓ Glossy Ibis
- ✓ White Ibis
- Scarlet Ibis
- ✓ Roseate Spoonbill

FLAMINGOS
- American Flamingo

WATERFOWL
- Canada Goose
- Blue Goose
- Fulvous Tree Duck
- ✓ Mallard
- ✓ Mottled Duck
- Gadwall
- Pintail
- Green-winged Teal
- ✓ Blue-winged Teal
- American Widgeon
- ✓ Shoveler
- Wood Duck
- Redhead
- Ring-necked Duck
- Canvasback
- Lesser Scaup
- Bufflehead
- Surf Scoter
- Ruddy Duck
- Masked Duck
- Hooded Merganser
- ✓ Red-breasted Merganser

VULTURES, HAWKS, & FALCONS
- ✓ Turkey Vulture
- ✓ Black Vulture
- ✓ Swallow-tailed Kite
- Everglade Kite
- Sharp-shinned Hawk
- Cooper's Hawk
- Red-tailed Hawk
- ✓ Red-shouldered Hawk
- Broad-winged Hawk
- Swainson's Hawk
- Short-tailed Hawk
- ✓ Bald Eagle
- ✓ Marsh Hawk
- ✓ Osprey
- ✓ Caracara
- Peregrine Falcon
- Pigeon Hawk
- ✓ Sparrow Hawk

GALLINACEOUS BIRDS
- ✓ Bobwhite
- Turkey

CRANES, RAILS, & ALLIES
- ✓ Sandhill Crane
- ✓ Limpkin
- King Rail
- Clapper Rail
- Virginia Rail
- ✓ Sora
- Yellow Rail
- Black Rail
- ✓ Purple Gallinule
- ✓ Common Gallinule
- ✓ American Coot

SHOREBIRDS, GULLS, & ALLIES
- Amer. Oystercatcher
- ✓ Semipalmated Plover
- Piping Plover
- Snowy Plover
- Wilson's Plover
- ✓ Killdeer
- American Golden Plover
- ✓ Black-bellied Plover
- ✓ Ruddy Turnstone
- ✓ Common Snipe
- Long-billed Curlew
- Whimbrel
- Upland Plover
- ✓ Spotted Sandpiper
- Solitary Sandpiper
- ✓ Willet
- ✓ Grtr. Yelw.-legs
- Lesr. Yelw.-legs
- Knot
- Purple Sandpiper
- Pectoral Sandpiper
- Whte-rumped Sandpiper
- Baird's Sandpiper
- Least Sandpiper
- Dunlin
- ✓ Sht-billed Dowitcher
- Long-billed Dowitcher
- Stilt Sandpiper
- ✓ Semipalm. Sandpiper
- Western Sandpiper
- ✓ Marbled Godwit
- ✓ Sanderling
- American Avocet
- Black-necked Stilt
- Wilson's Phalarope
- Pomarine Jaeger
- Parasitic Jaeger
- Grt. Blk-bckd. Gull
- ✓ Herring Gull
- ✓ Ring-billed Gull
- ✓ Laughing Gull
- Bonaparte's Gull
- Gull-billed Tern
- Forster's Tern
- Common Tern
- Roseate Tern
- Sooty Tern
- Bridled Tern
- Least Tern
- ✓ Royal Tern
- Sandwich Tern
- ✓ Caspian Tern
- Black Tern
- Brown Noddy Tern
- Black Noddy Tern
- ✓ Black Skimmer
- Dovekie

PIGEONS & DOVES
- White-crowned
- ✓ Rock
- White-winged
- ✓ Mourning
- ✓ Ground

CUCKOOS
- Mangrove Cuckoo
- Yelw.-billed Cuckoo
- Blck.-billed Cuckoo
- ✓ Smooth-billed Ani

OWLS
- Barn
- Screech
- Great Horned
- ✓ Burrowing
- Barred
- Short-eared

GOATSUCKERS
- Chuck-will's-widow
- Whip-poor-will
- Common Nighthawk

SWIFTS & HUMMINGBIRDS
- Chimney Swift
- ✓ Ruby-thrt. Hummingbird

KINGFISHERS
- ✓ Belted Kingfisher

WOODPECKERS
- ✓ Yelw.-shafted Flicker
- ✓ Pileated Woodpecker
- ✓ Red-bellied Woodpecker
- Red-headed Woodpecker
- ✓ Yelw.-bellied Sapsucker
- Hairy Woodpecker
- Downy Woodpecker
- Red-cockaded Woodpecker

FLYCATCHERS
- ✓ Eastern Kingbird
- Gray Kingbird
- Western Kingbird
- Scissor-tailed Flyctchr.
- ✓ Grt.-crstd. Flyctchr.
- ✓ Eastern Phoebe
- *Empidonax* flyctchrs.
- ✓ E. Wood Pewee
- Olive-sided Flyctchr.
- Vermillion Flyctchr.

SWALLOWS
- ✓ Tree Swallow
- Bank Swallow
- Rgh.-winged Swallow
- Barn Swallow
- Cliff Swallow
- ✓ Purple Martin

JAYS & CROWS
- ✓ Blue Jay
- Scrub Jay
- ✓ Common Crow
- ✓ Fish Crow

TITMICE
- Tufted Titmouse

NUTHATCHES
- Brwn.-headed Nuthatch

BULBULS
- Red-whiskered Bulbul

WRENS
- House
- Winter
- Carolina
- Long-billed Marsh
- ✓ Short-billed Marsh

MOCKINGBIRDS
- ✓ Mockingbird
- ✓ Catbird
- Brown Thrasher

THRUSHES
- ✓ Robin
- Wood Thrush
- Hermit Thrush
- Swainson's Thrush
- Grey-cheeked Thrush
- Veery
- E. Bluebird

GNATCATCHERS & KINGLETS
- Blue-gray Gnatcatcher
- Ruby-crowned Kinglet

PIPITS
- Water Pipit

WAXWINGS
- Cedar

SHRIKES
- ✓ Loggerhead

STARLINGS
- ✓ Starling

VIREOS
- ✓ White-eyed
- Bell's
- Yellow-throated
- Solitary
- Black-whiskered
- ✓ Red-eyed

Appendix Three: Bird Guides (with Associated Lifers)

Brian Bell (Ancient Murrelet)
Chris Benesh (Tropical Parula)
Gavin Bieber (White-tailed Ptarmigan)
Chris Brown (Himalayan Snowcock)
Rich Cimino (Mountain Quail)
Dave Currie (Pink-footed Goose)
Forrest Davis (Bristle-thighed Curlew)
Jeff Dingle (Caracara)
Jon Dunn (California Gnatcatcher)
Pete Dunne (Prothonotary Warbler)
Brian Gibbons (Blue-footed Booby)
Laurens Halsey (Sinaloa Wren)
Steve Heinl (Spectacled Eider)
Denver Holt (Great Gray Owl)
Alvaro Jaramillo (Marbled Murrelet)
Aaron Lang (Siberian Accentor)
Paul Lehman (Murphy's Petrel)
Larry Manfredi (Purple Swamphen)
Tom Mason (Allen's Hummingbird)
Michael O'Brien (Black Noddy)
Brian Patteson (Trindade Petrel)
Roger Tory Peterson (Yellow-headed Blackbird)
Bryan Pfeiffer (Mourning Warbler)
Jay Pitocchelli (Snowy Owl)
Robert Rugg (Blackburnian Warbler)
Leslie Severinghaus (Red-whiskered Bulbul)
Win Shafer (Tundra Bean-Goose)
Debra Shearwater (Laysan Albatross)
David Sibley (Northern Jacana)
Dave Stejskal (Buff-collared Nightjar)
Mandy Talpas (Mariana Swiftlet)
Bill Thompson III (Sprague's Pipit)
Jeff Wahl (Kirtland's Warbler)
Chris Wood (Black Rosy-Finch)
Howard Wood (Gyrfalcon)

Appendix Four: Mega-Rarities in Taxonomic Order

Code-5s (10)

Zenaida Dove
Whiskered Tern
Bat Falcon
Great Black Hawk
Amazon Kingfisher
Social Flycatcher
Sinaloa Wren
Blue Mockingbird
Red-legged Thrush
Black-backed Oriole

Code-4s (30)

Pink-footed Goose
Barnacle Goose
White-cheeked Pintail
Garganey
Oriental Turtle-Dove
Key West Quail-Dove
Common Crane
Northern Lapwing
Northern Jacana
White-winged Tern
Blue-footed Booby
Little Egret
Steller's Sea-Eagle
Tufted Flycatcher
Thick-billed Vireo
Tamaulipas Crow
Middendorff's Grasshopper-Warbler
Common Chiffchaff
Dusky Warbler
Bahama Mockingbird
Red-flanked Bluetail
Siberian Accentor
Common Chaffinch
Little Bunting
Streak-backed Oriole
Golden-crowned Warbler
Slate-throated Redstart
Crimson-collared Grosbeak
Yellow Grosbeak
Black-faced Grassquit

Code-3s (41)

Tundra Bean-Goose
Tufted Duck
American Flamingo
Ruddy Ground-Dove
Smooth-billed Ani
Common Cuckoo
Buff-collared Nightjar
White-eared Hummingbird
Black-tailed Godwit
Ruff
Sharp-tailed Sandpiper
Curlew Sandpiper
Red-necked Stint
Gray-tailed Tattler
Common Greenshank
Ivory Gull

Ross's Gull
Slaty-backed Gull
Arctic Loon
Trindade Petrel
Murphy's Petrel
Flesh-footed Shearwater
Hook-billed Kite
Ferruginous Pygmy-Owl
Aplomado Falcon
Rose-throated Becard
LaSagra's Flycatcher
Fork-tailed Flycatcher
Black-capped Gnatcatcher
Rufous-backed Robin
White Wagtail
Red-throated Pipit
Brambling
Rustic Bunting
Five-striped Sparrow
Western Spindalis
Shiny Cowbird
Tropical Parula
Rufous-capped Warbler
Flame-colored Tanager
Morelet's Seedeater

Appendix Five: Top 20 Target Species

Whooper Swan
Smew
Black Swift
Plain-capped Starthroat
Berylline Hummingbird
Yellow Rail
Great Skua
Craveri's Murrelet
Whiskered Auklet
Red-billed Tropicbird
White-tailed Eagle
Eared Quetzal
Yellow-green Vireo
Fieldfare
Redwing
Aztec Thrush
Hawfinch
McKay's Bunting
Bananaquit
Blue Bunting

About the Author

George C. Wood is a fundraiser for his high school alma mater, The Haverford School, located outside of Philadelphia. A 1979 graduate of Princeton University with an M.S. in environmental science from The Ohio State University (1995), George is married to Nina Weisbord. A proud father of six children and two grandchildren, he currently resides in Haverford, Pennsylvania and plays many racquet sports when not birding.

Made in the USA
Monee, IL
11 November 2022

2a257691-a397-489b-b2d9-619a3b649399R01